Mathematics Education for Students with Learning Disabilities

MATHEMATICS EDUCATION

FOR STUDENTS WITH

LEARNING DISABILITIES

THEORY TO PRACTICE

EDITED BY

Diane Pedrotty Rivera

8700 Shoal Creek Boulevard
Austin, Texas 78757-6897

This text has been adapted from material that previously appeared in the *Journal of Learning Disabilities.*

pro·ed

© 1998 by PRO-ED, Inc.
8700 Shoal Creek Boulevard
Austin, Texas 78757-6897

Library of Congress Cataloging-in-Publication Data

Mathematics education for students with learning disabilities : theory
   to practice / edited by Diane Pedrotty Rivera.
        p.   cm.
   Includes bibliographical references and indexes.
   ISBN 0-89079-710-2 (alk. paper)
   1. Mathematics—Study and teaching—United States.   2. Learning
disabled children—Education—United States.   I. Rivera, Diane
   Pedrotty.
QA13.M144   1998   97-38019
510'.71—dc21   CIP

Printed in the United States of America
1   2   3   4   5   6   7   8   9   10   02   01   00   99   98

Contents

v

Foreword

MARGO A. MASTROPIERI

It is truly exciting to see such an innovative and comprehensive volume on mathematics and learning disabilities (LD) become available for those of us interested in these areas. This book is indisputably the first scholarly collection addressing all aspects of mathematics and LD, including characteristics and assessment of mathematics LD, mathematics programming and interventions, and teacher preparation. In addition, this volume provides a historical perspective on educational reform efforts in mathematics and discusses the implications of these reforms for the education of students with LD. The authors of several chapters not only address the recent reforms, they also discuss the relationship of these ideas to current efforts in special education research. I am especially impressed with the comprehensiveness of, and intellectual approach to, the topics in this book.

Mathematics is a critically important but understudied area in learning disabilities research. Indeed, studies have indicated that 64% of students with LD are performing below grade level in mathematics (McLesky & Waldron, 1990). Moreover, as Rivera has noted in her introductory chapter, minimal research has been conducted on the relationship of LD and mathematics as compared to other areas, such as reading. For example, in their 1989 research, Lessen, Dudzinski, Karsh, and Van Acker identified only a few published intervention studies of students with LD, and even fewer intervention studies (22) in mathematics. More recent reviews of mathematics research for individuals with LD noted an increase in the number of studies published and provided analyses of types of interventions (Jitendra & Xin, 1997; Mastropieri, Scruggs, & Shiah, 1991).

The dominant research focus in special education historically has been on computation, problem-solving strategies using specifically taught procedural steps, teacher-directed instruction, and drill-and-practice formats (Mastropieri et al., 1991). Although these approaches have been found to be successful in increasing student learning, they are in direct opposition to ideas promoted in recent reform efforts by the National Council of Teachers of Mathematics (NCTM, 1989; National Research Council, 1989). Both the NCTM standards and reports from the National Research Council encouraged conceptual knowledge, student-generated constructivist learning, and deep understanding of mathematics concepts rather than knowledge of basic facts, rules, and procedures. Although these reports addressed the needs of general education students and mathematics, the needs of individuals with disabilities—and how they fit within this framework—were not adequately covered. This book makes a significant contribution toward addressing this gap. It not only offers information pertinent to both positions, but also integrates the two seemingly very diverse views into a comprehensible "state of the art" for education in mathematics for individuals with LD. Such an understanding and integration of curriculum reform efforts in general education and intervention efforts in special education is essential.

One of the many strengths of this volume is the organizational framework presented by Rivera in her opening chapter. She eloquently provides the context for the entire volume by presenting a historical overview of the field of mathematics, summaries of current reform efforts in mathematics and research in mathematics and LD, and a discussion of the major issues. She sets the stage for bridging the gap between mathematics reform and LD by providing brief descriptions of the ensuing chapters. It becomes immediately evident that Rivera has assembled some of the most distinguished experts in the country on mathematics and LD, who, in turn, have made significant contributions to this work. The result of this combination of theoretical perspective research and issues in characteristics, assessment, programming, and teacher preparation is a remarkable volume that offers the very best in current knowledge from general education mathematics, mathematics and LD, and teacher preparation. It can be considered a handbook both for practices in mathematics and LD—past and present—and for determining the future research agenda in this very important area. I do not doubt that this book will become essential reading for current and future special educators.

1. Mathematics Education and Students with Learning Disabilities

DIANE PEDROTTY RIVERA

The field of mathematics special education has experienced significant growth in response to an increased awareness of the prevalence of mathematics learning disabilities (LD) and the pervasive and persistent nature of the ensuing difficulties across the elementary and secondary school curricula (Cawley & Miller, 1989; Garnett, 1987). Furthermore, results from follow-up and follow-along studies indicate that individuals with mathematics learning disabilities tend not to perform at a level commensurate with their peers' on the basic functional skills (e.g., telling time, counting change) that are necessary for successful adult living (Wagner, 1990).

A growing body of research has produced important findings about characteristics of mathematics learning disabilities, assessment, and mathematics programming. For instance, investigations have shown that students may exhibit difficulties using effective cognitive and metacognitive problem-solving strategies (Montague & Applegate, 1993), memory and retrieval processes (Bley & Thornton, 1995), and generalization skills (Woodward, 1991), which stem from developmental delays (Cawley & Miller, 1989; Ginsberg, 1989) and cognitive deficits. These difficulties are manifested in the inability to acquire and apply mathematical skills and concepts, to reason, and to solve mathematical problems. The resulting poor mathematics performance (ranging from mild to severe) in traditional curricula impedes students' abilities to compete with their typical peers (Cawley, Baker-Kroczynski, & Urban, 1992). Varying levels of support are

1

needed to accommodate individual student needs, which in some cases means intensive, individualized special education instruction.

In response to these chronic mathematical difficulties, special education researchers have identified critical instructional and curricular variables that promote effective programming for students with mathematics learning disabilities (Carnine, chapter 6 of this book; Cawley & Parmar, 1992; Mercer & Miller, 1992; Rivera & Smith, 1987). Drawing from cognitive and developmental psychology and behavioral theory, the knowledge base of what constitutes sensible instructional practices for youngsters with math LD has increased significantly in recent years. For instance, researchers have identified discrete instructional design variables, such as explicit strategies (Kelly, Gersten, & Carnine, 1990; Montague & Bos, 1986), relevant practice (Jones, Wilson, & Bhojwani, chapter 8 of this book), and alternative algorithms (Cawley & Parmar, 1992), that foster mathematical understanding and evaluative thinking and promote a better match between learners with math difficulties and the skills and concepts presented for instruction.

Assessment practices for documenting the existence of mathematics LD and student progress in the curriculum have evolved to embrace norm-referenced instruments, criterion-referenced assessment, and certain nonstandardized procedures, such as clinical interviews, error analysis, and portfolios (Bryant & Rivera, chapter 5 of this book; Hammill, 1987). The focus is not only the mathematical products students generate, but also the processes and conceptual understanding they use to arrive at solutions.

Research findings regarding characteristics of mathematics learning disabilities, special education mathematics instructional programming, and effective assessment practices have ramifications for teacher preparation programs in learning disabilities. Teacher educators must conceptualize all the domains of teachers' knowledge and identify ways to integrate math course content taught by their general education colleagues and special education methodology into a cohesive teacher preparation program.

This book on mathematics education and students with learning disabilities is devoted to a synthesis of special education research on perspectives of mathematics learning disabilities, assessment practices, best practices in mathematics programming, and teacher preparation issues. These chapters provide important information about the difficulties students with mathematics learning disabilities encounter in mastering mathematics curricula, and about the implications for programmatic decision making at the elementary and secondary levels. The authors offer a wealth of ideas for professionals who teach children, conduct mathematics research, or prepare teachers at the preservice and inservice levels.

This introductory chapter provides an overview of trends in the fields of mathematics education and mathematics special education; these trends are discussed in greater detail by the other authors. This chapter is divided into two major sections. The first focuses on the field of mathematics education. Because mathematics special education is directly linked to the field of mathematics education, and because students with LD receive most of their education in the general education classroom, special educators must be knowledgeable about trends in the field of mathematics education and remain aware of possible instructional ramifications of those trends for students with learning disabilities. The second section presents an overview of the four major topics discussed in the book: perspectives on math LD, assessment, mathematics programming, and teacher preparation. The chapter concludes with a focus on future directions in mathematics special education.

THE FIELD OF MATHEMATICS EDUCATION

This section presents an overview of the evolution of the field of mathematics education from 1950 to the present. The year 1950 was chosen as a starting point because the 1950s, along with the 1960s, were viewed as the "golden age" of mathematics education and thus seemed like a good place to begin a historical discussion.

The evolution of the field of mathematics education has been shaped by sociopolitical factors (e.g., performance scores of U.S. youth on national mathematics achievement tests [Lindquist, Carpenter, Silver, & Mathews, 1983; McKnight et al., 1987], the discretion of textbook publishers, international events, professional organizations) and specific trends (e.g., developments in scientific methodologies of inquiry, theoretical explanations of how children learn and come to understand mathematics, and developments in the field of psychometrics and assessment). The following is an overview of some of the sociopolitical forces and trends that have contributed to the development of the field of mathematics education. The trends include reform, theory and instruction, and research methodologies. The information has been organized by decades and is intended only to provide readers with an overview of the evolution of the field; naturally, some overlap of decades and trends occurs (see the *Journal for Research in Mathematics Education,* Vol. 25, No. 6, for a more thorough description).

Mathematics Reform

The 1950s and 1960s. Kilpatrick (1992) labeled the 1950s and 1960s "the golden age" in the field of mathematics education because of the prolific national support for research in mathematics instruction and learning during that time. The '50s and '60s were marked by a surge of federal funding supportive of mathematics research and the development of a specialized field that was intended to produce scholars, teacher educators, highly trained mathematics teachers, and graduates who could succeed in competitive markets to reinstate the United States as a world leader. Spurred by the emergence of the Soviets' advancement in the space program (e.g., Sputnik) and consumer groups' criticism of public school students' math knowledge (leveled at schools officials), funding for math program development and expansion of graduate programs was substantial during this time period (e.g., the Cooperative Research Act, the National Longitudinal Study of Mathematical Abilities funded by the National Science Foundation).

The 1970s and 1980s. Three significant reform efforts occurred in the 1980s: the "back to basics" movement, increased math-course credit graduation requirements, and minimum competency testing. These efforts were sparked by national commission reports on the content, teaching, and standards in public schools (e.g., *A Nation at Risk* [National Commission on Excellence in Education, 1983]; *Educating Americans for the 21st Century* [National Science Board Commission, 1983]; *Making the Grade* [Twentieth Century Fund, 1983) and were linked to trends occurring in both the 1970s and 1980s. For instance, the 1970s were marked by disillusionment with mathematics instructional approaches (e.g., discovery learning, heuristic methods of problem solving) predicated on research lacking in theoretical and empirical bases (Kulm, cited in Kieran, 1994). Additionally, national test scores indicative of recurring poor student mathematics achievement (e.g., in fractions, decimals, story problems) plagued this time period (Carpenter, Coburn, Reyes, & Wilson, 1976; Carpenter, Corbitt, Kepner, Lindquist, & Reyes, 1981; Dossey, Mullis, Lindquist, & Chambers, 1988).

Perhaps one of the most significant reform efforts of the 1980s was instituted by the National Council of Teachers of Mathematics (NCTM; Lester, 1994). Their publications, *An Agenda for Action* (1980) and the *Curriculum and Evaluation Standards for School Mathematics* (or the Standards, 1989), initiated change in teacher preparation programs and public school mathematics curricula and instruction. Professionals were interested in preparing students for the demands of living in the 21st century by (a) broadening the scope of mathematics curricula to include such topics as estimation, problem solving, and geometry; and (b) reducing the emphasis of "traditional" curricula consisting primarily of basic computational skills. These documents would generate debate and reform in teacher preparation programs and public school curricula and instruction over the course of the next decade.

The 1990s. The field of mathematics education has undergone reform efforts aimed at improving mathematics instruction for all students. As a result of technological advancements (Office of Technology Assessment, 1988), national movements (e.g., Goals 2000: The Educate America Act), and mathematical workforce expectations (Johnston & Packers, 1987), mathematics professionals have set forth recommendations for restructuring instructional and curricular emphases in mathematics programs (e.g., National Council of Supervisors of Mathematics, 1988; National Council of Teachers of Mathematics [NCTM], 1989).

The goal of the mathematics reform movement is to enable all students to develop a strong mathematics education, which, the organizations assert, is "at the basis of the nation's need for a competent work force and an informed society" (Conference Board of the Mathematical Sciences, 1995, p. 1). NCTM has led the way in advocating change in mathematics curricula, instruction, assessment, and teacher preparation programs. At the heart of the reform movement is a paradigmatic shift from predominantly reductionistic, skills-based instruction to a constructivist epistemology that encompasses active student learning rooted in problem-solving situations facilitated by teachers' guidance and questioning.

In particular, the Standards have generated discussion and controversy in the professional community. The Standards consist of statements specifying topics for three levels: Grades K through 4, Grades 5 through 8, and Grades 9 through 12 (see Table 1.1 for a sample of the Standards across the three levels). The Standards were established to provide a framework for guiding mathematics curricula and pedagogy and for promoting more activity-based inquiry in mathematics teaching.

Table 1.1

NCTM Standards

Grades K–4

Standard 1: Mathematics as Problem Solving
- Use problem-solving approaches to investigate and understand mathematical content.
- Formulate problems from everyday and mathematical situations.
- Develop and apply strategies to solve a wide variety of problems.
- Verify and interpret results with respect to the original problem.
- Acquire confidence in using mathematics meaningfully.

Standard 2: Mathematics as Communication
- Relate physical materials, pictures, and diagrams to mathematical ideas.
- Reflect on and clarify their thinking about mathematical ideas and situations.
- Relate their everyday language to mathematical language and symbols.
- Realize that representing, discussing, reading, writing, and listening to mathematics are a vital part of learning and using mathematics.

Standard 3: Mathematics as Reasoning
- Draw logical conclusions about mathematics.
- Use models, known facts, properties, and relationships to explain their thinking.
- Justify their answers and solution processes.
- Use patterns and relationships to analyze mathematical situations.
- Believe that mathematics makes sense.

(table continues)

(Table 1.1 continued)

Standard 4: Mathematical Connections
- Link conceptual and procedural knowledge.
- Relate various representations of concepts or procedures to one another.
- Recognize relationships among different topics in mathematics.
- Use mathematics in other curriculum areas.
- Use mathematics in their daily lives.

Standard 5: Estimation
- Explore estimation strategies.
- Recognize when an estimate is appropriate.
- Determine the reasonableness of results.
- Apply estimation in working with quantities, measurement, computation, and problem solving.

Standard 6: Number Sense and Numeration
- Construct number meanings through real-world experiences and the use of physical materials.
- Understand our numeration system by relating counting, grouping, and place-value concepts.
- Develop number sense.
- Interpret the multiple uses of numbers encountered in the real world.

Standard 7: Concepts of Whole Number Operations
- Develop meaning for the operations by modeling and discussing a rich variety of problem situations.
- Relate the mathematical language and symbolism of operations to problem situations and informal language.
- Recognize that a wide variety of problem structures can be represented by a single operation.
- Develop operation sense.

Standard 8: Whole Number Computation
- Model, explain, and develop reasonable proficiency with basic facts and algorithms.
- Use a variety of mental computation and estimation techniques.
- Use calculators in appropriate computational situations.
- Select and use computation techniques appropriate to specific problems and determine whether the results are reasonable.

Standard 9: Geometry and Spatial Sense
- Describe, model, draw, and classify shapes.
- Investigate and predict the results of combining, subdividing, and changing shapes.
- Develop spatial sense.
- Relate geometric ideas to number and measurement ideas.
- Recognize and appreciate geometry in their world.

Standard 10: Measurement
- Understand the attributes of length, capacity, weight, mass, area, volume, time, temperature, and angle.
- Develop the process of measuring and concepts related to units of measurement.
- Make and use estimates of measurement.
- Make and use measurements in problem and everyday situations.

Standard 11: Statistics and Probability
- Collect, organize, and describe data.
- Construct, read, and interpret displays of data.
- Formulate and solve problems that involve collecting and analyzing data.
- Explore concepts of chance.

(table continues)

(Table 1.1 continued)

Standard 12: Fractions and Decimals
- Develop concepts of fractions, mixed numbers, and decimals.
- Develop number sense for fractions and decimals.
- Use models to relate fractions to decimals and to find equivalent fractions.
- Use models to explore operations on fractions and decimals.
- Apply fractions and decimals to problem situations.

Standard 13: Patterns and Relationships
- Recognize, describe, extend, and create a wide variety of patterns.
- Represent and describe mathematical relationships.
- Explore the use of variables and open sentences to express relationships.

Grades 5–8

Standard 1: Mathematics as Problem Solving
- Use problem-solving approaches to investigate and understand mathematical concepts.
- Formulate problems from situations within and outside mathematics.
- Develop and apply a variety of strategies to solve problems, with emphasis on multi-step and nonroutine problems.
- Verify and interpret results with respect to the original problem situation.
- Generalize solutions and strategies to new problem situations.
- Acquire confidence in using mathematics meaningfully.

Standard 2: Mathematics as Communication
- Model situations using oral, written, concrete, pictorial, graphical, and algebraic methods.
- Reflect on and clarify their own thinking about mathematical ideas and situations.
- Develop common understandings of mathematical ideas, including the role of definitions.
- Use the skills of reading, listening, and viewing to interpret and evaluate mathematical ideas.
- Discuss mathematical ideas and make conjectures and convincing arguments.
- Appreciate the value of mathematical notation and its role in the development of mathematical ideas.

Standard 3: Mathematics as Reasoning
- Recognize and apply deductive and inductive reasoning.
- Understand and apply reasoning processes, with special attention to spatial reasoning and reasoning with proportions and graphs.
- Make and evaluate mathematical conjectures and arguments.
- Validate their own thinking.
- Appreciate the pervasive use and power of reasoning as a part of mathematics.

Standard 4: Mathematical Connections
- See mathematics as an integrated whole.
- Explore problems and describe results using graphical, numerical, physical, algebraic, and verbal mathematical models or representations.
- Use a mathematical idea to further their understanding of other mathematical ideas.
- Apply mathematical thinking and modeling to solve problems that arise in other disciplines, such as art, music, psychology, science, and business.
- Value the role of mathematics in our culture and society.

Standard 5: Number and Number Relationships
- Understand, represent, and use numbers in a variety of equivalent forms (integer, fraction, decimal, percentage, exponential, and scientific notation) in real-world and mathematical problems.

(table continues)

(Table 1.1 continued)

- Develop number sense for whole numbers, fractions, decimals, integers, and rational numbers.
- Understand and apply ratios, proportions, percentages in a wide variety of situations.
- Investigate relationships among fractions, decimals, and percentages.
- Represent numerical relationships in one- and two-dimensional graphs.

Standard 6: Number Systems and Number Theory
- Understand and appreciate the need for numbers beyond the whole numbers.
- Develop and use order relations for whole numbers, fractions, decimals, integers, and rational numbers.
- Extend their understanding of whole number operations to fractions, decimals, integers, and rational numbers.
- Understand how the basic arithmetic operations are related to one another.
- Develop and apply number theory concepts (e.g., primes, factors, and multiples) in real-world and mathematical problem situations.

Standard 7: Computation and Estimation
- Compute with whole numbers, fractions, decimals, integers, and rational numbers.
- Develop, analyze, and explain procedures for computation and techniques for estimation.
- Develop, analyze, and explain methods for solving proportions.
- Select and use an appropriate method for computing from among mental computation, paper-and-pencil, calculator, and computer methods.
- Use computation, estimation, and proportions to solve problems.
- Use estimation to check the reasonableness of results.

Standard 8: Pafferns and Functions
- Describe, extend, analyze, and create a wide variety of patterns.
- Describe and represent relationships with tables, graphs, and rules.
- Analyze functional relationships to explain how a change in one quantity results in a change in another.
- Use patterns and functions to represent and solve problems.

Standard 9: Algebra
- Understand the concepts of variable, expression, and equation.
- Represent situations and number patterns with tables, graphs, verbal rules, and equations and explore the interrelationships of these representations.
- Analyze tables and graphs to identify properties and relationships.
- Develop confidence in solving linear equations using concrete, informal, and formal methods.
- Investigate inequalities and nonlinear equations informally.
- Apply algebraic methods to solve a variety of real-world and mathematical problems.

Standard 10: Statistics
- Systematically collect, organize, and describe data.
- Construct, read, and interpret tables, charts, and graphs.
- Make inferences and convincing arguments that are based on data analysis.
- Evaluate arguments that are based on data analysis
- Develop an appreciation for statistical methods as powerful means for decision making.

Standard 11: Probability
- Model situations by devising and carrying out experiments or simulations to determine probabilities.
- Model situations by constructing a sample space to determine probabilities.
- Appreciate the power of using a probability model by comparing experimental results with mathematical expectations.

(table continues)

(Table 1.1 continued)

- Make predictions that are based on experimental or theoretical probabilities.
- Develop an appreciation for the pervasive use of probability in the real world.

Standard 12: Geometry
- Identify, describe, compare, and classify geometric figures.
- Visualize and represent geometric figures with special attention to developing spatial sense.
- Explore transformations of geometric figures.
- Represent and solve problems using geometric models.
- Understand and apply geometric properties and relationships.
- Develop an appreciation of geometry as a means of describing the physical world.

Standard 13: Measurement
- Extend their understanding of the process of measurement.
- Estimate, make, and use measurements to describe and compare phenomena.
- Select appropriate units and tools to measure to the degree of accuracy required in a particular situation.

Grades 9–12

Standard 1: Mathematics as Problem Solving
- Use, with increasing confidence, problem-solving approaches to investigate and understand mathematical content.
- Apply integrated mathematical problem-solving strategies to solve problems from within and outside mathematics.
- Recognize and formulate problems from situations within and outside mathematics.
- Apply the process of mathematical modeling to real-world problem situations.

Standard 2: Mathematics as Communication
- Reflect upon and clarify their thinking about mathematical ideas and relationships.
- Formulate mathematical definitions and express generalizations discovered through investigations.
- Express mathematical ideas orally and in writing.
- Read written presentations of mathematics with understanding.
- Ask clarifying and extending questions related to mathematics they have read or heard about.
- Appreciate the economy, power, and elegance of mathematical notation and its role in the development of mathematical ideas.

Standard 3: Mathematics as Reasoning
- Make and test conjectures.
- Formulate counterexamples.
- Follow logical arguments.
- Judge the validity of arguments.
- Construct simple valid arguments.
      and so that, in addition, college-intending students can—
      Construct proofs for mathematical assertions, including indirect proofs and proofs by mathematical induction.

Standard 4: Mathematical Connections
- Recognize equivalent representations of the same concepts.
- Relate procedures in one representation to procedures in an equivalent representation.
- Use and value connections among mathematical topics.
- Use and value the connections between mathematics and other disciplines.

(table continues)

(Table 1.1 continued)

Standard 5: Algebra

- Represent situations that involve variable quantities with expressions, equations, inequalities, and matrices.
- Use tables and graphs as tools to interpret expressions, equations, and inequalities.
- Operate on expressions and matrices, and solve equations and inequalities.
- Appreciate the power of mathematical abstraction and symbolism.
  and so that, in addition, college-intending students can—
  Use matrices to solve linear systems.
- Demonstrate technical facility with algebraic transformations, including techniques based on the theory of equations.

Standard 6: Functions

- Model real-world phenomena with a variety of functions.
- Represent and analyze relationships using tables, verbal rules, equations, and graphs.
- Translate among tabular, symbolic, and graphical representations of functions.
- Recognize that a variety of problem situations can be modeled by the same type of function.
- Analyze the effects of parameter changes on the graphs of functions.
  and so that, in addition, college-intending students can—
  Understand operations on, and the general properties and behavior of, classes of functions.

Standard 7: Geometry From a Synthetic Perspective

- Interpret and draw three-dimensional objects.
- Represent problem situations with geometric models and apply properties of figures.
- Classify figures in terms of congruence and similarity and apply these relationships.
- Deduce properties of, and relationships between, figures from given assumptions.
  and so that, in addition, college-intending students can—
  Develop an understanding of an axiomatic system through investigating and comparing various geometries.

Standard 8: Geometry From an Algebraic Perspective

- Translate between synthetic and coordinate representations.
- Deduce properties of figures using transformations and using coordinates.
- Identify congruent and similar figures using transformations.
- Analyze properties of Euclidean transformations and relate translations to vectors.
  and so that, in addition, college-intending students can—
- Deduce properties of figures using vectors.
- Apply transformations, coordinates, and vectors in problem solving.

Standard 9: Trigonometry

- Apply trigonometry to problem situations involving triangles.
- Explore periodic real-world phenomena using the sine and cosine functions.
  and so that, in addition, college-intending students can—
  Understand the connection between trigonometric and circular functions.
- Use circular functions to model periodic real-world phenomena.
- Apply general graphing techniques to trigonometric functions.
- Solve trigonometric equations and verify trigonometric identities.
- Understand the connections between trigonometric functions and polar coordinates, complex numbers, and series.

Standard 10: Statistics

- Construct and draw inferences from charts, tables, and graphs that summarize data from real-world situations
- Use curve fitting to predict from data.

(table continues)

Table 1.1 continued)

- Understand and apply measures of central tendency, variability, and correlation.
- Understand sampling and recognize its role in statistical claims.
- Design a statistical experiment to study a problem, conduct the experiment, and interpret and communicate the outcomes.
- Analyze the effects of data transformations on measures of central tendency and variability.
  and so that, in addition, college-intending students can—
  Transform data to aid in data interpretation and prediction.
- Test hypotheses using appropriate statistics.

Standard 11: Probability

- Use experimental or theoretical probability, as appropriate, to represent and solve problems involving uncertainty.
- Use simulations to estimate probabilities.
- Understand the concept of random variable.
- Create and interpret discrete probability distributions.
- Describe in general terms the normal curve and use its properties to answer questions about sets of data that are assumed to be normally distributed.
  and so that, in addition, college-intending students can—
  Apply the concept of a random variable to generate and interpret probability distributions including binomial, uniform, normal, and chi square.

Standard 12: Discrete Mathematics

- Represent problem situations using discrete structures such as finite graphs, matrices, sequences, and recurrence relations.
- Represent and analyze finite graphs using matrices.
- Develop and analyze algorithms.
- Solve enumeration and finite probability problems.
  and so that, in addition, college-intending students can—
  Represent and solve problems using linear programming and difference equations.
- Investigate problem situations that arise in connection with computer validation and the application of algorithms.

Standard 13: Conceptual Underpinnings of Calculus

- Determine maximum and minimum points of a graph and interpret the results in problem situations.
- Investigate limiting processes by examining infinite sequences and series and areas under curves.
  and so that, in addition, college-intending students can—
  Understand the conceptual foundations of limit, the area under a curve, the rate of change, and the slope of a tangent line, and their applications in other disciplines.
- Analyze the graphs of polynomial, rational, radical, and transcendental functions.

Standard 14: Mathematical Structure

- Compare and contrast the real number system and its various subsystems with regard to their structural characteristics.
- Understand the logic of algebraic procedures.
- Appreciate that seemingly different mathematical systems may be essentially the same.
  and so that, in addition, college-intending students can—
  Develop the complex number system and demonstrate facility with its operation.
- Prove elementary theorems within various mathematical structures, such as groups and fields.
- Develop an understanding of the nature and purpose of axiomatic systems.

Although intuitively appealing to many professionals, the purpose and content of the Standards have been greeted with mixed reviews. Many special education professionals acknowledge the importance of increasing the emphasis on mathematics problem solving and including content that traditionally has received little attention as part of a total mathematics program (e.g., estimation, probability, statistics, geometry). However, the Standards also were challenged by special educators (e.g., Hofmeister, 1993; Hutchinson, 1993a; Mercer, Harris, & Miller, 1993; Rivera, 1993), who identified the following issues and made the accompanying recommendations:

1. *Limited reference to students with disabilities.* Although the Standards refer to "all students" and to students "who have not been successful in school," students with learning disabilities are not explicitly mentioned. It is recommended that discussions centered on the implementation of the Standards include decisions about ways to address the individualized needs of students with mathematics learning disabilities.
2. *Vague theoretical constructs* (i.e., general problem solving, inquiry/discovery learning, math reasoning). Implementation of instructional practices based on ill-defined theoretical constructs raises efficacy questions and ghosts from special education past. Researchers can approach this issue by defining constructs, determining the validity of those constructs, and conducting research that documents their instructional worth.
3. *Limited replicable, validated instructional practices.* A detailed description of the research base underlying the recommendations for curricular and pedagogical reform is important. There is a need for empirical validation that includes well-defined target populations (including students with learning disabilities), treatments, instrumentation, and results.
4. *Shift in paradigmatic thinking.* The Standards embrace a constructivist approach to teaching children mathematics and to explaining how children come to understand mathematical concepts. Although the merits of constructivism for teaching youngsters with LD have been recognized in the special education literature (e.g., Poplin, Wiest, & Thorson, 1995; Thornton, Langrall, & Jones, chapter 7 of this book), a wealth of research-validated instructional practices for students with LD (e.g., Cawley et al., 1992; Gersten, Carnine, & Woodward, 1987; Mercer & Miller, 1992; Montague & Bos, 1986; Rivera & Smith, 1987) must be maintained as part of effective special education mathematics programming (see the discussion on research-validated practices in the second part of this chapter).

As the field of mathematics education continues to evolve, special education researchers and practitioners must remain players in the reform process by engaging in dialogue and classroom-based research with their mathematics education colleagues. At issue is the translation of reform recommendations into valid instructional programs that are sensitive to the characteristics of students with mathematics disabilities and that promote mathematics achievement.

Theory and Instruction

The 1950s and 1960s. Mathematics education became a specialized field of study in the 1950s and 1960s. As was true with other educational fields, math education was inspired by the works of Skinner (1968), Gagné (1962), and Bloom (1956) and by the widespread acceptance of behaviorism as an explanation for learning. During that time, math instructional activities were based more on intuition than on scientific research methodologies (Scandura, cited in Johnson, Romberg, & Scandura, 1994). However, at the demand of math consumers and federal funding agents, such practices gave way to objectives-based formative evaluations of mathematics programs (Scandura, cited in Johnson et al., 1994).

This time period (and into other decades) is also characterized by the influences of Piaget's (1954, 1963) work grounded in cognitive developmental psychology. His contributions stemmed from assumptions about how children develop ideas through accommodation and assimilation, and from observations about the role of the environment in shaping those ideas (Kieran, 1994). Finally, the 1960s, in particular, were marked by training projects focusing on pedagogical approaches that were rooted in the "discovery teaching" methodology and the School Mathematics Study Group textbooks, which were influential in setting a direction for mathematics curricula nationwide (Cooney, 1994).

The 1970s and 1980s. In the 1970s and 1980s, researchers in the field of mathematics continued to search for theoretical models that could explain the phenomena of children's learning and understanding of mathematics. This was a transition time wherein math researchers were breaking with behavioral tradition and seeking cognitive models as frameworks for mathematics education (Kieran, 1994).

Behaviorism and the work of Gagné and Bloom continued to influence mathematics instruction, during the 1970s in particular. The conceptualization of students' mathematical understanding was directly linked to behavioral theory. Instruction was viewed as a taxonomy of math educational objectives consisting of prerequisite skills, and the learning of mathematics was related to students' abilities to retain, transfer, and recall mathematical information as measured by standardized tests.

Also emergent in the 1970s was the influence of cognitive psychology in conceptualizing children's understanding of mathematics and defining the "learning" of mathematics. Piagetian (1963, 1970) learning theory of children's cognitive processes and the theoretical writings of Skemp (1976), Vygotsky (1978), and Wittrock (1974) facilitated a shift (a) from viewing mathematics instruction as manipulating situations (à la stimulus–response behavioral tradition) to fostering learning within a social context (Kieran, 1994), and (b) from assessing mathematical knowledge solely in terms of standardized, performance-based measures to examining thinking processes through nonstandardized procedures (e.g., "thinking aloud," error analysis).

These theoretical shifts sparked debate about mathematics instruction as "skills versus understanding," and discussion about the role of the teacher and classroom in mathematics learning. Investigations (e.g., Fennema, Carpenter, & Lamon, 1988) were undertaken to explore cognitively based instruction emphasizing how children process and construct mathematical knowledge, the situations that foster mathematical understanding, and the role of the teacher in promoting this understanding.

Particularly noteworthy during the 1980s was the emphasis on mathematical problem solving evident in the NCTM (1980, 1989) publications, the comprehensive reviews of problem-solving research (e.g., Lester & Garofalo, 1982; Schoenfeld, 1985), and the research-based recommendations for classroom instruction (e.g., Driscoll, 1983; Suydam, 1980). According to Lester (1994), of particular concern during this time period were issues pertaining to problem-solving "task variables" (Goldin & McClintock, 1984), distinctions between "expert" and "novice" problem solvers (Schoenfeld, 1985), instructional practices and programs that promoted effective problem solving, and the relationship between metacognitive and cognitive behaviors associated with problem solving (Schoenfeld, 1982). The shift to investigations of children's mathematical cognition affected assessment practices as educators sought practices that captured students' use of cognitive strategies to learn content (Bryant & Rivera, chapter 5 of this book; Schoenfeld, 1994).

The 1990s. The field of mathematics education is witnessing phenomenally accelerated growth, as evidenced by the burgeoning research on instructional methodology and on how children come to understand mathematical concepts (Schoenfeld, 1994). Shaped by the theoretical writings of Vygotsky (1978) and information-processing theory, mathematics educators have widely embraced the notion that learning mathematics is a constructive process (Kieran, 1994). "Knowing" mathematics and constructing mathematical knowledge is viewed as an ongoing social activity within a dynamic classroom culture, influenced by the teacher (Cobb, Yackel, & Wood, 1992).

Research Methodologies

The 1950s Through the 1970s. The topic of research methodologies is considered an important component of the evolution of the field of mathematics because it reflects the philosophic and theoretical evolution of how mathematics researchers viewed the nature of mathematics understanding and instruction. From the 1950s to about the 1970s, research methodology and questions of inquiry, grounded in behaviorism, reflected an empirical, analytic tradition whereby researchers conducted efficacy studies on instructional methodologies (e.g., math problem solving, discovery learning). Investigators studied the effects of math programs on students' achievement and the attainment of math concepts at each level of Bloom's taxonomy (Kieran, 1994); however, lacking in such research were descriptions of the kinds of teaching procedures used, the processes of mathematics learning that were part of the daily routine, and the instructional materials used in the treatment condition (Kieran, 1994). Also, research methodologies were plagued by poorly defined constructs (e.g., heuristic problem solving, discovery learning), limited descriptions of treatment conditions, and virtually no discussion about how students' understanding of mathematics evolved; interpretation and generalization of results were limited at best.

The 1980s to the Present. In the 1980s and 1990s, a broader range of research methodologies has been embraced by mathematics researchers to capture more fully the nature of understanding mathematics (Schoenfeld, 1994). Case studies and ethnographic descriptions as well as more traditional scientific methods of inquiry are pervasive in the mathematics research literature as researchers attempt to provide rich, descriptive accounts of how children learn mathematics. Social settings and classroom environments are now recognized as important factors contributing to how children develop mathematical cognition. As theories of mathematics learning and instruction evolve, so do the research methodologies used to study various phenomena.

The first section of this chapter provided an overview of the evolution of the field of mathematics education since the 1950s. The overview framed research and practice in the field of mathematics special education. The remaining section describes information pertinent to the field of mathematics special education and the four topic areas of this book.

THE FIELD OF MATHEMATICS SPECIAL EDUCATION

This book focuses on content relevant to four topic areas: perspectives on mathematics learning disabilities, assessment, mathematics programming, and teacher preparation. Each of the four topic areas is described with (a) an overview of the topic, (b) summaries of the chapters pertinent to that topic, and (c) emergent themes stemming from those chapters.

Perspectives on Mathematics Learning Disabilities

A significant number of students with learning disabilities exhibit problems with mathematics (Scheid, 1990). For instance, in one survey, teachers reported that 26% of their students with LD received services primarily for mathematical difficulties encountered in the general education curriculum (McLeod & Armstrong, 1982). Researchers have noted that mathematics learning disabilities are evident throughout the school years, and that students' mathematics performance tends to plateau at the fifth- or sixth-grade level (Cawley et al., 1992; Cawley & Miller, 1989). Moreover, mathematics deficiencies continue to plague adults with learning disabilities in postsecondary settings (Adelman & Vogel, 1991) and in everyday living activities (Patton, Cronin, Bassett, & Koppel, chapter 10 of this book).

Researchers have documented specific mathematical deficiencies, primarily in the areas of computation and word problems. In the area of computation, students may exhibit deficits and limited proficiency related to fact retrieval (Garnett & Fleischner, 1983), problem conceptualization, speed of processing (Geary, 1993), and use of effective calculation strategies (e.g., counting on, counting all, counting fingers; Geary, 1990; Goldman, Pellegrino, & Mertz, 1988). Although some of these children may experience delays in acquiring procedural knowledge strategies, others exhibit developmental differences (e.g., retrieval deficits) that do not necessarily disappear with age (Geary, 1993).

Solving word problems tends to be problematic for many students in general and for students with mathematics deficiencies in particular. Studies have linked difficulties with solving word problems to sentence structure complexity, presence of extraneous information, semantic complexity, cognitive and metacognitive deficiencies, reading problems, poor computational skills, the problem's representation, and difficulty with multiple steps (Englert, Culatta, & Horn, 1987; Parmar, 1992; Parmar, Cawley, & Frazita, 1996). In particular, when studying the word problem cognitive and metacognitive strategy skills of middle school students with LD, Montague (chapter 9 of this book) did not find quantitative differences in strategy use among ability groups but did find that students demonstrating poor problem solving ability differed quantitatively and qualitatively in *types* of strategies used, as compared with their typical peer group.

Although much has been learned about the mathematical skills of students with LD, research is needed in other math areas, such as place value, geometry, fractions, measurement, and so forth. Some research has been initiated in algebra (Hutchinson, 1993b); however, a great deal of work remains to be done in developing a better understanding of specific difficulties in other math concepts and skills.

In this book, three perspectives on mathematics learning disabilities are discussed: developmental, neurological and neuropsychological, and educational. Research conducted in these three areas has contributed significantly to our understanding of mathematics learning disabilities. The following is a brief introduction to each perspective.

Developmental. Herbert P. Ginsburg (chapter 2) presents a thoughtful discussion of how we might approach an examination of mathematics learning disabilities from a developmental perspective. He offers an important discussion on informal and formal mathematics learning, describing how typical children, regardless of cultural heritage, come to understand the fundamentals of informal mathematics (beginning long before formal schooling occurs) through their interactions with physical and social environments that are "rich in quantitative information." From these environmental interactions and linguistic representations of mathematical experiences (e.g., *more, less, greater than*), children construct their own interpretations and understandings of mathematics as they develop quantitative concepts. Inspired by curiosity and motivated by "practical utility," children "invent" counting strategies and develop notions about mathematical quantity and concepts.

Ginsburg provides readers with a distinction between informal and formal (school) mathematics and presents an ecological description of schools with inherent problems in mathematics education. He discusses issues pertinent to mathematics learning disabilities, including the challenges professionals face in distinguishing children with mathematics learning disabilities from those who possess "garden-variety" poor mathematics achievement. He also emphasizes the importance of framing math cognitive difficulties within "situated contexts" sensitive to settings, topics of instruction, and children's ages. He concludes with a discussion of methods that may tap children's cognitive processes and reveal important developmental and instructional information about how children with math LD come to understand mathematics.

Neurological and Neuropsychological. Byron P. Rourke and James A. Conway (chapter 3) present a thorough discussion of the historical and conceptual roots of arithmetic disabilities from a neurological and neuropsychological perspective. As is true in other areas of learning disabilities, the conceptual beginnings of dyscalculia can be traced to historically "rich and descriptive" case studies of adult patients who sustained neurological damage manifested in various types of arithmetical disorders (i.e., acquired acalculia). Findings from extensive error analyses led researchers (e.g., Hécaen, Angelergues, & Houillier, 1961) to posit neurological substrates (or subtypes) of calculations based on "presumed neurological mechanisms underlying each type." These findings contributed to the conceptualization of brain–behavior relationships in calculation and served as a framework for the identification of developmental disorders of arithmetic. Although the work of Hécaen et al. was significant, Rourke and Conway acknowledge the difficulties involved in generalizing clinical findings from adult patients to children.

Rourke and Conway present a pertinent discussion about developmental dyscalculia, acknowledging the need for an acceptable definition that can guide research and diagnosis. They highlight the work of Kosc (1974), who provided one definition of dyscalculia and identified six arithmetical subtypes. The final section of their chapter deals with a neurodevelopmental approach to arithmetic disorders, reporting research findings on two subtypes of arithmetic disabilities. The authors present their results on neuropsychological profiles pertaining to different patterns of academic achievement and the potential of neuropsychological assessment to predict academic performance.

Educational. Susan Peterson Miller and Cecil D. Mercer (chapter 4) analyze educational factors that contribute to mathematics learning disabilities and, consequently, poor mathematics performance, and offer recommendations for resolving some of those problems. National reform movements, learner characteristics, and curricula and instruction are cited as factors that cause or exacerbate mathematics learning disabilities.

Reform efforts have dominated the field of mathematics over the years. As educators respond to curricular and instructional changes, so too must students for whom mathematics is difficult at best. Miller and Mercer provide a thorough discussion of learner characteristics related to mathematics performance. Drawing from cognitive psychology research, they provide descriptions of math problems based on information processing theory and cognitive and metacognitive processing deficits. Poorly designed instructional methodologies and curricular materials are cited as significant factors contributing to math difficulties. Issues such as textbook instruction, accommodations for individual mathematics needs in general education classrooms, and limited field-testing of math materials are highlighted.

Miller and Mercer offer practical, thoughtful recommendations for approaching educational factors that may impede mathematics performance. In particular, they argue for mathematics "refinement" rather than continual reform. Additionally, they discuss ways to implement effective accommodations in classrooms and provide recommendations for best practices in mathematics education.

Themes. First, mathematics learning disabilities have historically received less attention than reading disorders. However, that trend has shifted as special educators have come to recognize the prevalence of mathematics learning disabilities. Research studies have revealed that students experience difficulties in acquiring and understanding mathematical concepts and techniques.

Second, students with mathematics learning disabilities exhibit complex learning and behavioral characteristics that can be traced to developmental, neurological, neuropsychological, and educational influences. Interpretations of mathematics learning disabilities within these four areas provide an extensive understanding of how they are exacerbated by certain curricular and instructional demands.

Finally, the role of research in further identifying and explaining mathematics learning disabilities is vital. More information is needed to better understand children's construction of mathematical knowledge, the different types of mathematical disorders, and the effects of classroom demands (e.g., textbooks, instruction) on math learning.

Assessment

Assessment has been defined as "the systematic process of gathering educationally relevant information to make legal and instructional decisions about the provision of special education services" (McLoughlin & Lewis, 1990, p. 4). The purpose of assessment is to seek answers to specific questions (e.g., Is there a school performance problem? What are the student's academic strengths and weaknesses? What strategies does he or she use to solve problems? Does the student have a learning disability in mathematics?) to assist professionals and parents in making sound educational decisions about students' individual needs. The formal assessment process (i.e., administration of diagnostic standardized tests to identify strengths, weaknesses, and the presence of a learning disability) has been controversial because of issues surrounding (a) overidentification of students from culturally and linguistically diverse backgrounds as having disabilities, (b) difficulties in discriminating between low achievement and learning disabilities (Epps,

Ysseldyke, & McGue, 1984), (c) statistical procedures for determining a severe discrepancy (Dangel & Ensminger, 1988; Reynolds, 1985), and (d) bias of assessment information (Frame, Clarizio, & Porter, 1984).

Opponents of traditional testing support the adoption of "alternative" approaches to assessment, including portfolio assessments, performance-based assessments, and authentic assessments (Feuer & Fulton, 1993; Wesson & King, 1992; Worthern, 1993). Linked to school reform, these alternative assessment approaches are popular as a result of interest in (a) examining student achievement and progress more directly; (b) focusing on analyzing processes rather than just products; (c) identifying assessment options to traditional, standardized testing procedures; and (d) measuring higher order thinking and problem-solving skills rather than basic skills alone.

Efforts to reform mathematics curricula and assessment (e.g., NCTM, 1989; National Research Council, 1989) have focused on redefining mathematics curricula and assessment procedures, which examine student performance in relation to instructional goals and techniques. Mathematics educators are encouraged to select assessment procedures that (a) reflect the school's curriculum; (b) use instructional materials (e.g., manipulatives, calculators); (c) include different response formats (e.g., pencil-and-paper, verbal, manipulative); and (d) compare the ability of particular students with that of their peer group. Table 1.2 provides examples of assessment purposes and procedures that can be used to answer assessment questions. In particular, it is important to note that the Standards support a variety of assessment practices (e.g., norm-referenced, criterion-referenced, nonstandardized), depending on the purpose of assessment.

Rivera, Taylor, and Bryant (1994–95) reviewed the special education literature since 1990 to identify studies of mathematics assessment practices that were aligned with the NCTM Standards' call for more curriculum-based assessment. Rivera et al. identified studies of mathematics assessment procedures that assisted teachers in (a) identifying math learning problems, (b) planning math instruction, and/or (c) evaluating program effectiveness. Findings revealed empirical studies validating curriculum-based measurement (e.g., Fuchs, Fuchs, Hamlett, & Stecker, 1990; Fuchs, Fuchs, Hamlett, & Stecker, 1991; Phillips, Hamlett, Fuchs, & Fuchs, 1993) for improving instructional decision making and evaluating student progress, and protocol analysis (Parmar, 1992) for examining the cognitive processing difficulties of students engaged in problem solving. Thus we see that special educators embrace various assessment procedures (as outlined in Table 1.2) to answer specific assessment questions relating to students' educational needs and progress.

In this book, Brian R. Bryant and Diane Pedrotty Rivera (chapter 5) present an extensive discussion of the evolution of mathematics assessment,

Table 1.2
Math Assessment Purposes, Questions, and Methods

Purposes and questions	Methods
Identification and eligibility	
• Is there a mathematics performance problem? • Is the problem related to a disability? • How does the student's performance compare to that of his or her peer group? • What are the student's overall strengths and weaknesses in mathematics? • What is the student's current level of performance in the math concepts and skills?	• Norm-referenced assessment instruments in mathematics • Criterion-referenced instruments • Standardized achievement tests
Instructional diagnosis	
• Does this student possess the prerequisite skills necessary for new learning? • What does this student comprehend about the concept or skill? • What strategies does this student use to solve problems? • What accounts for this student's not generalizing previously taught material to newly presented material? • Does the student require more practice?	• Norm-referenced assessment instruments in mathematics • Criterion-referenced instruments • Curriculum-based measurement • Error analysis (nonstandardized) • "Think aloud" cognitive problem solving (nonstandardized)
Program evaluation	
• Has the student achieved mastery? • Is the instructional intervention effective? • What modifications and reteaching are necessary?	• Norm-referenced assessment instruments in mathematics • Criterion-referenced instruments • Curriculum-based measurement

covering the Pioneering, Proliferation, and Refinement periods of assessment development. The chronology highlights societal influences on the assessment field as test developers and schools have responded to criticisms leveled at test construction and the reform movements that have been pervasive in public school education.

The Pioneering Period was characterized by the development of arithmetic tests, laying the foundation for test administration and scoring guidelines. Test results were used to identify arithmetic teaching trends.

The Proliferation Period saw the rise of test popularity as psychologists and educators identified assessment practices for identifying faulty computational strategies and suggesting remedial techniques. During this time, achievement and intelligence tests were used as comparative measures.

The Refinement Period was characterized by professional organizations' close scrutiny of the technical merits of test instruments and their subsequent endorsement of more stringent test development. The field of learning disabilities emerged during the Refinement Period, thus necessitating the development of psychometrically sound mathematics instruments that could facilitate the identification of specific learning disabilities.

Bryant and Rivera provide a description of various assessment strategies that assist practitioners in identifying students' curricular strengths and weaknesses and cognitive processing deficits in solving mathematical problems. These strategies are in line with the recommendation by the NCTM (1989) that educators use techniques that offer insight into how children *understand* mathematics. Of particular importance is that student progress is monitored regularly.

Themes. First, the technical adequacy of test instruments remains a current issue in the assessment field. Teachers must ensure that assessment tools are used only for the purposes for which they were intended.

Second, assessment practices are sensitive to sociopolitical influences (e.g., recommendations by professional organizations, reform movements) and emergent theoretical orientations (e.g., constructivism, information processing).

Third, various types of assessment techniques (e.g., norm-referenced, criterion-referenced) are relevant when assessing students with mathematics disabilities. The assessment questions to be answered drive the selection of the technique.

Mathematics Programming: Curricula and Instruction

As reform efforts shape the field of mathematics special education, and as research reveals additional theoretical findings about mathematics learning disabilities, researchers and practitioners must ensure that effective curricular and instructional practices for students with math deficiencies are identified and implemented. In recent years, two special education literature reviews (Lessen, Dudzinski, Karsh, & Van Acker, 1989; Mastropieri, Scruggs, & Shiah, 1991) were conducted revealing empirically based instructional interventions that can be used to enhance the mathematics performance of students with learning disabilities.

Lessen et al. (1989) identified 135 academic (e.g., reading, mathematics, spelling) intervention studies conducted from 1978 to 1987. It is interesting that only 16% of the total studies were concerned with mathematics instruction, especially computation (19, computation; 1, time-telling; 2, word problems). Mastropieri et al.'s (1991) literature review (1975–1988) on mathematics instruction identified intervention research that they categorized according to theoretical approaches to instruction and alternative delivery systems of instruction (peer mediation and technology). The Mastropieri et al. study included a broader spectrum of journals and a longer time span than the Lessen et al. study and drew from *Dissertation Abstracts International*; 30 intervention studies were identified. Findings from Mastropieri et al. reflect (a) the influence of theoretical orientations, (b) a continued emphasis on computational research, (c) research in story-problem solving that parallels problem-solving reform efforts by NCTM, and (d) the role technology plays in mathematics instruction. The intervention studies identified in the two reviews are shown in Table 1.3 and are organized by theoretical approach, technology, and student-mediated arrangements; a sample of other intervention research conducted since 1988 also is included.

The information in Table 1.3 shows certain trends in mathematics special education intervention research. First, relative to other academic areas (particularly reading), mathematics has received limited intervention research emphasis; in recent years, that trend appears to be changing. Second, the influences of emergent theoretical approaches in general and special education have been pervasive in mathematics intervention research. The information in Table 1.3 shows that researchers have employed principles of cognitive psychology as well as behavioral theory in their investigations of techniques that assist students in learning, primarily, how to solve computation and word problems. The role of technology in mathematics instruction is as an area that requires considerable research. Also, as students with LD spend more time in general education settings, various student-mediated approaches (e.g., peer tutoring, cooperative learning) must be examined to determine their usefulness in fostering mathematics performance. Table 1.3 also illustrates that special education mathematics intervention research lacks examination of techniques that teach students a variety of mathematical concepts. This trend may change as special education researchers explore techniques from a constructivist approach.

As Cawley and Parmar (1992) noted, special educators are at a crossroads regarding program development for students with math LD in respect to addressing the NCTM's call for curricular reform and promoting students' understanding and mastery of mathematics. This book reflects an interest in examining math learning from different perspectives (e.g., developmental/cognitive, neuropsychological, behavioral, constructivist). Of utmost importance is that special educators heed calls for refinement, maintain empirically sound instructional practices, and continue investigations into variables that constitute best practice.

In this book, the topic area of mathematics programming includes five chapters that discuss instructional design, programming for elementary students, practices for secondary students, story-problem–solving research, and instruction from a life-span viewpoint.

Table 1.3
Trends in Mathematics Interventions

Researchers	Skills	Interventions
Cognitive approaches		
Lloyd, Saltzman, & Kauffman, 1981	Basic facts	Count-bys
Leon & Pepe, 1983	Computation	Self-instructions
Jones, Thornton, & Toohey, 1985	Basic facts	Counting strategies
Marzola, 1985	Word problems	Six-step problem-solving strategy
Schunk, 1985	Computation	Goal setting
Montague & Bos, 1986	Word problems	Eight-step cognitive strategy
Schunk & Cox, 1986	Computation	Verbalization and feedback
Case & Harris, 1988	Word problems	Strategy & self-instruction
Collins & Carnine, 1988	Problem solving	Learner-verification strategy; CAI
Fuchs, Bahr, & Rieth, 1989	Computation	Goal setting: assigned versus self-selected
McIntyre, Test, Cook, & Beattie, 1991	Basic facts	Count-bys
Case, H arris, & Graham, 1992	Word problems	Self-instructional strategy
Mercer & Miller, 1992	Basic facts and word problems	Concrete–semiconcrete–abstract; FAST DRAW strategy
Montague, 1992	Word problems	Cognitive & metacognitive strategy
Harding, Gust, Goldhawk, & Bierman, 1993	Computation	Interactive unit
Hutchinson, 1993	Algebra	Cognitive strategy
Miller & Mercer, 1993	Basic facts	SOLVE strategy
Behavioral approaches		
Smith & Lovvitt, 1976	Computation	Reinforcement
Smith & Lovitt, 1975	Computation	Demonstration + permanent model
Rivera & Smith, 1987		
Rivera & Smith, 1988		
Blankenship, 1978	Computation	Modeling, demonstration, and feedback
Blankenship & Baumgartner, 1982		
Luiselli & Downing, 1980	Computation	Reinforcement and feedback
Peterson, Mercer, & O'Shea, 1988	Place value	Concrete–semiconcrete–abstract teaching sequence
Cybriwsky & Schuster, 1990	Basic facts	Constant time delay
Kelly, Gersten, & Carnine, 1990	Factions	Range of examples
Koscinski & Gast, 1993	Basic facts	Constant time delay
Zawaiza & Gerber, 1993	Word problems	Explicit instruction
Technology approaches		
Friedman & Hofmeister, 1984	Telling time	Microcomputer–videodisc instruction
Trifiletti, Frith, & Armstrong, 1984	Math skills	CAI/resource room
Chiang, 1986	Computation	CAI
Kelly, Carnine, Gersten, & Grossen, 1986	Fractions	Videodisc instruction/traditional textbook instruction

(table continues)

(Table 1.3 continued)

Researchers	Skills	Interventions
Carnine, Englemann, Hofmeister, & Kelly, 1987	Fractions	Videodisc instruction
Howell, Sidorenko, & Jurica, 1987	Basic facts	Drill and practice vs. tutorial-based software
Fuchs, Fuchs, & Hamlett, 1989	Computation	Computers; corrective feedback
Gleason, Carnine, & Boriero, 1990	Word problems	CAI
Bottge & Hasselbring, 1993	Word problems	Videodisc instruction
Rivera, Carter, & Smith, 1996	Basic facts	Computer-based instruction vs. fluency-building techniques
Student-mediated approaches		
Kane & Alley, 1980	Computation	Peer tutoring
Slavin, Madden & Leavey, 1984	Computation	Team-assisted individualization
Maheady, Sacca, & Harper, 1987	Computation	Cooperative learning
Beirne-Smith, 1991	Computation	Peer tutoring

Note. CAI = computer-assisted instruction.

Instructional Design. Douglas Carnine (chapter 6) presents readers with a thorough discussion of instructional design variables critical for effective mathematics instruction. He attributes poor mathematics performance to a "mismatch" between the design of teaching procedures and curricular materials and the learning characteristics associated with learning disabilities. Carnine poses an important question for researchers and practitioners regarding the challenges of (a) ensuring that students with mathematics learning disabilities acquire the skills necessary for successful mathematics performance, and (b) offering instruction that engages students in activities designated by the Standards.

Carnine identifies five major components of effective instructional design: "big ideas," conspicuous strategies, efficient use of time, explicit instruction, and procedures that promote retention. Evident in these components are research-based techniques for fostering the generalization and integration of skills.

Programming for Elementary- and Secondary-Level Students. Carol A. Thornton, Cynthia W. Langrall, and Graham A. Jones (chapter 7) present the elementary-level perspective of mathematics instruction by arguing for instruction that encompasses four major themes viewed as promising practices by math educators for teaching students with mathematics learning disabilities: (a) providing a broad and balanced mathematics curriculum;

(b) engaging students in rich, meaningful problem activities; (c) accommodating the diverse ways in which children learn; and (d) encouraging students to discuss and justify their strategies and solutions.

Thornton et al.'s work is grounded in a constructivist orientation and emphasizes a mathematics reasoning and problem-solving approach to instruction. The authors present findings from four studies involving students with mathematics learning disabilities, followed by case illustrations of the four themes. Evident in the findings is an emphasis on students' collaborative interactions to resolve mathematical problems, student-directed instruction, and an inquiry approach to instruction coupled with individual modifications.

Eric D. Jones, Rich Wilson, and Shalini Bhojwani (chapter 8) offer the secondary-level perspective of mathematics instruction. The authors note that research in secondary-level mathematics is limited in scope compared to research conducted with younger children; however, some of the elementary findings in the special education literature are applicable to the secondary level. Jones et al. offer a comprehensive discussion of the variables that confound efforts to promote effective instruction at the secondary level. They offer a thorough discussion of issues related to prior achievement, content of secondary-level instruction, management of instruction, and evaluation of instruction. Their work offers important insights into the connections among self-efficacy, performance, and effective instruction, which they contend can be strengthened by effective principles of instructional design.

They describe three important curricular factors that contribute to the quality of instruction and cite effective teacher- and peer-mediated instructional arrangements to foster mathematics performance. It is important to note that they make a strong case for data-based instruction linked to the curriculum and related to instructional decision making and program modifications.

Jones et al. conclude their chapter with a discussion about the belief systems currently governing mathematics education. They highlight underlying assumptions of constructivism and discuss three possible obstacles to effective implementation of constructivist practices.

Problem Solving. Marjorie Montague (chapter 9) offers a thorough discussion of the teaching of story-problem–solving skills to students with mathematics learning disabilities. She provides a theoretical and research base for cognitive strategy instruction grounded in a developmental perspective of mathematics learning disabilities.

Three major sections are featured in her chapter. First, Montague presents a cogent discussion of theoretical bases supporting cognitive strategy instruction for mathematical problem solving, which has been a line of research for her for over a decade. She traces the influence of information processing theory, including the metacognitive aspects, on problem-solving activities.

Second, Montague discusses, from a developmental perspective, the cognitive characteristics that may interfere with students' abilities to acquire types of strategies, and notes the qualitative and quantitative differences in types of strategies used by students with mathematics learning disabilities compared with their typical peer group. She argues for explicit problem-solving strategies—guided learning in problem-solving instruction rather than an emphasis on textbook and "real-world" story problems. Finally, she reviews a body of research on domain-specific cognitive strategy instruction conducted with middle school students with LD, noting improved academic achievement; increased strategic knowledge; and procedures for assessment, instruction, and evaluation in problem solving.

Life Skills. James R. Patton, Mary E. Cronin, Diane S. Bassett, and Annie E. Koppel (chapter 10) provide an important discussion about math life skills, which are critical for the successful transition to adulthood by students with learning disabilities. The chapter is divided into four sections that focus on curricular and instructional considerations for teaching life skills. In the first section, the authors provide a foundation by describing life skills and presenting a rationale for their inclusion in classroom instruction. In particular, the authors make a strong case for incorporating life skills instruction throughout the K through 12 curriculum.

In the second section, the authors link life skills instruction to current reform efforts by citing the relationship of life skills curricula to the specific goals and content of the Standards and Goals 2000. There is a natural connection between life skills curricula and the math Standards (e.g., estimation, problem solving, measurement, etc.). The third section features a discussion of the mathematical skills that adults need in their daily lives. The authors describe various settings (home, work, postsecondary) in which mastery of basic mathematical skills is essential. In the final section, the authors present practical instructional ideas for integrating life skills into the curriculum and offering subject area course work to prepare students for the demands of adulthood.

Themes. First, curricular organization and instructional delivery are critical variables in effective programming for students with mathematics learning disabilities. Instruction must be tailored to address individual needs, modified accordingly, and evaluated carefully to ensure that learning is occurring.

Second, the principles of effective instructional design, direct instruction, cognitive strategy training, and constructivist practices contribute to students' learning and understanding of mathematics.

Third, mathematics problem solving is a critical life skill that should be taught across the grades. Students can learn to use effective, explicit cognitive and metacognitive strategies for solving mathematical word problems. Students benefit from engaging in a variety of activities in which they have opportunities to discuss and justify their solution strategies.

Fourth, students with mathematics learning disabilities benefit from teacher- and peer-mediated instruction. Peer-mediated instruction is beneficial when teachers provide guidance, prompts, and feedback.

Teacher Preparation

Lappan and Theule-Lubienski (cited in Cooney, 1994) conceptualized teacher preparation programs in mathematics as consisting of three domains of knowledge: mathematics, pedagogy of mathematics, and students. Lappan and Theule-Lubienski called for teacher educators to identify the information and beliefs that constitute the domains, and the preparatory experiences teachers need to acquire this body of knowledge (Cooney, 1994).

Coupled with the three domains identified by Lappan and Theule-Lubienski must be a body of knowledge special educators should possess to work effectively with youngsters with learning disabilities. The challenge for special education teacher educators is to work collaboratively with their mathematics education colleagues and state department officials to construct a sound teacher preparation program so that future teachers graduate with competencies not only in general mathematics education but also in their field of special education.

To assist teacher educators in identifying important competencies for future teachers, the Council for Exceptional Children issued a special education competency list, and the Division for Learning Disabilities published their *DLD Competencies for Teachers of Students with Learning Disabilities* (Graves et al., 1992). The conceptual basis of the DLD competencies includes general education preparation, a knowledge-base dimension, a service delivery dimension, and competency categories consisting of 209 competencies and clinical field experience. Mathematics fall within the "Specialized Instructional Strategies, Technologies, and Materials" category and contain 12 competencies covering topics such as curricula, instructional techniques, remedial techniques, readiness, developmental teaching progressions, mathematics programming, and research.

In this book, Rene S. Parmar and John F. Cawley (chapter 11) present an important discussion regarding issues about, and techniques for, preparing teachers to teach mathematics to students with learning disabilities. They use the *Professional Standards for Teaching Mathematics* (NCTM, 1991) and the DLD competencies to frame their case for examining current preparation practices and changing the way we prepare our future special education teachers. Of critical importance is whether teachers are being prepared to challenge their students mathematically (i.e., incorporate the Standards into teaching), to teach their students important mathematical life skills, and to use instructional techniques that deviate from the more traditional paper-and-pencil, rote-memorization instruction.

Parmar and Cawley emphasize the importance of modeling good mathematics teaching in university course work through hands-on problem-solving activities. They stress the need for teachers to have a sound knowledge base in mathematics, including breadth and depth in curriculum and instruction. They remind us how crucial it is for future teachers to understand their students' specific mathematical difficulties; such awareness enables them to plan effective mathematics programs and to be well grounded in mathematics pedagogy. Parmar and Cawley acknowledge that an individual requires time to develop as a "teacher of mathematics" and that he or she should become familiar with different theoretical orientations to the teaching and learning of mathematics. They conclude by offering insight into how teachers can continue to develop professionally.

Themes. First, teacher preparation programs should be based on competencies designated by professional organizations. Second, teacher educators should model effective teaching practices as part of their delivery of instruction in college courses. Third, future teachers should receive specific course work instruction in mathematics curricula and pedagogy. Fourth, future teachers must develop a good understanding of mathematics learning disabilities to help them plan appropriate mathematics programs. Fifth, professional development is ongoing and should be continued throughout one's teaching career.

Future Directions

The final chapter in this book was written by Susan R. Goldman and Ted S. Hasselbring. They conclude by offering their perspectives on the ideas contained in the chapters and their insights into directions in which the field is heading. Additional research must be conducted in the four topic areas discussed in this book; perceptive reflection on how this research might evolve is left to Goldman and Hasselbring.

SUMMARY

The prevalence of students with mathematics learning disabilities has triggered an interest among special education researchers and practitioners in developing an understanding of the needs of this group of students, and in identifying effective instructional programming to foster their mathematical performance during the school years and into adulthood. Research into the characteristics of students with mathematics learning disabilities is being approached from different perspectives, including developmental, neurological and neuropsychological, and educational. This diversity helps us develop a broader understanding of students' learning needs and difficulties.

Special education assessment practices encompass a variety of approaches, including norm-referenced, criterion-referenced, and nonstandardized procedures, depending on the specific assessment questions professionals seek to answer. Students' mathematical knowledge and conceptual understanding must be examined to determine their strengths and weaknesses, curriculum-based progress, and use of cognitive strategies to arrive at mathematical solutions.

Research findings have identified empirically validated interventions for teaching mathematics curricula to students with mathematics learning disabilities. Research studies have been grounded in behavioral theory and cognitive psychology, with an emergent interest in the constructivist approach. Although research studies have focused primarily on computational performance, more work is being conducted in the areas of story-problem solving and technology. These areas as well as other math curricular skills require further study. Additionally, the needs of adults with math LD have spurred educators to examine the elementary and secondary math curricula and determine ways to infuse them with life skills instruction accordingly.

As the field of mathematics special education continues to evolve, special educators must remain cognizant of the developments in and influences on the field of mathematics education. Reform efforts have shaped the field significantly since the 1950s, contributing to the curriculum offered in mathematics textbooks and the pedagogical practices taught in higher education courses. Mathematics educators continue to search for a better understanding of how children learn mathematics; this process is shaped by the prevailing theoretical orientations and research methodologies.

This book on mathematics special education provides readers with information about the characteristics of students with mathematics learning disabilities, assessment procedures, mathematics programming, teacher preparation, and future directions for the field. It originated as a result of discussions with Lee Wiederholt and Judith K. Voress, who saw a need for the compilation of recent research and best practices in mathematics special education. I thank them for their support of and thoughtful insights about the development of this book. I also appreciate the support of George Hynd and his editorial assistant, Kathryn Black, in finalizing the details for publication. Finally, I am most appreciative of the authors' contributions; their work continues to significantly influence the development of the field of mathematics special education and programming for students with mathematics learning disabilities.

AUTHOR'S NOTE

The author wishes to thank Judith K. Voress for her feedback of an earlier draft of this chapter.

2. Mathematics Learning Disabilities: A View From Developmental Psychology

HERBERT P. GINSBURG

Most research on learning disabilities has focused on difficulties in the area of reading, with the result being that little attention has been given to mathematics learning disabilities. Perhaps (in the United States at least) this results in part from our culture's general reluctance to deal with things mathematical. Just as elementary-school teachers tend to avoid mathematics at all costs, so perhaps do researchers shy away from examining mathematics learning disabilities. Whatever the reason for the lack of popularity of this kind of research, the topic of learning disabilities in mathematics requires serious attention. Many children receive diagnoses of mathematics learning disability, or the related dyscalculia and acalculia. Yet, little is understood concerning these "conditions" and how they develop.

My goal in this chapter is to show how a developmental perspective can help to provide insights into the nature, development, and treatment of mathematics learning disabilities. Over the past 20 years, a good deal of research has been conducted on the development of mathematical thinking, mostly in normally achieving, middle class children in the United States, but also in children from different cultures, including street children in Brazil (Nunes, Schliemann, & Carraher, 1993) and children from unschooled societies in Africa (Ginsburg, Posner, & Russell, 1981). Researchers have examined a wide range of topics, from infants' perception of quantitative differences (Antell & Keating, 1983) to college students' understanding of the notion of limit (Ferrini-Mundy & Lauten, 1993). In general, researchers in this tradition eschew notions of general mathematics achievement or aptitude (as well as group or individually administered tests of such), and instead engage in the detailed investigation of thought processes in individual children. The research they have produced is among the most exciting in developmental psychology: It shows how knowledge originally developed in the "natural environment" contributes to and becomes transformed by what is taught in school.

I believe that the developmental perspective and the resulting body of research on the development of mathematical thinking have a good deal to say about how we should think about mathematics learning disabilities. The perspective and the research can provide concepts to clarify the analysis of mathematics learning disabilities, to help researchers think productively about the origins of those disabilities and their remediation, to raise important questions for future investigations, and to suggest research strategies and methods. So, this article is about how we *might* approach the study of mathematics learning disabilities from a developmental perspective. It is not a review of the literature, not a compendium of what has already been discovered about mathematics learning disabilities (after all, the existing research results are sparse), although I do refer to previous research where it is relevant.

This chapter is organized in several sections. First, I consider the robust development of "informal mathematics" in young children, mainly outside of the school setting. Next, I describe what happens when ordinary children encounter ordinary schooling in the United States: Learning difficulties are inevitable in our imperfect system of education. All this then provides the foundation for an analysis of math learning disabilities.

INFORMAL MATHEMATICS

To understand children's intellectual development, researchers usually take an interactionist, constructivist point of view, which draws heavily on the work of both Piaget and Vygotsky. In this tradition, researchers begin by considering the development of informal knowledge in children's ordinary environments.

Natural Learners, Natural Environments

According to Piaget (1952b), children have a biologically based propensity to learn. They accommodate environmental demands, and they assimilate what the environment has to offer. Children are natural learners; they are intrinsically motivated. They learn because their minds are biologically designed to develop concepts and modes of thought useful for adaptation to the environment.

The mind always develops in an environment, both physical and social. The normal mind with a biologically based propensity for learning—this mind of children from every country and culture and social class and racial and ethnic group—develops in an environment that is in many respects rich in quantitative information and events. All children encounter small, discrete objects that can be manipulated, moved, arranged, touched, and counted. All children come across groups of objects that are more numerous than others; they encounter things larger and greater in volume than others. Moreover, the physical environment of quantity appears to offer rich stimulation across a wide variety of cultures. In what culture, however impoverished, do children lack things to count or add or compare? Such mathematical events and phenomena are pervasive and fundamental; indeed, they appear to be universal in the physical world.

Children also encounter a social environment that provides important mathematical experiences. They hear adults counting, or see them using money; they see numerals on telephones, VCRs, buses, houses, and television programs. In the physical and social environments, quantity and number are all around us; they cannot be avoided. As the philosopher Whitehead (1929) put it, "You cannot evade quantity. You may fly to poetry and music, and quantity and number will face you in your rhythms and octaves" (p. 7).

Of course, these environments of quantity vary considerably from culture to culture. American children can count blocks; Dioula children (in Africa) can count stones. Some cultures offer an everyday written mathematics—as on telephones or microwave ovens—and others do not. But almost all known cultures offer a fundamental mathematical system—namely, the counting words. Moreover, in virtually all cultures studied, counting systems are highly elaborate, most often involving a base 10 system and extending to rather large numbers (Dantzig, 1954). Furthermore, research has shown that some "primitive" cultures offer surprisingly rich mathematical environments. For example, the Zinacantan Indians of Mexico weave patterns rich in geometric complexity (Greenfield, 1984).

Human languages all contain means for describing quantitative events. Among a baby's first words are "more" and "another" (Bloom, 1970). Children's literature around the world offers folktales and stories involving elaborate quantitative notions (Smith & Wendelin, 1981). For example, the bears in "The Three Bears" vary in size according to age and gender: Baby is small, mama larger, and papa largest. The bears have beds, bowls, and chairs that also vary in size. Moreover, this variation is systematically related to (or positively correlated with) the size of the bears: Baby bear gets the smallest bowl, mama the bigger, and papa the biggest. Because books usually do not describe quantitative notions like these in explicit mathematical terms, the readers and the children read to are blissfully unaware that they must be involved in mathematical thinking to understand the story line.

In brief, throughout development, before and after entrance to school, children are normally exposed to physical and social environments rich in mathematical opportunities. Children encounter quantity in the physical world, counting numbers in the social world, and mathematical ideas in the world of story and literature.

What the Child Constructs

Children do not simply absorb information from the world; they are not simply shaped by the environment. Rather, children actively *construct* concepts, understandings, strategies, and modes of thought (Noddings, 1990). Of course, each child (and each adult) develops constructions that are to some extent "unique." Nevertheless, at a given developmental level, children's constructions are similar, and often different from adults' (see Note 1).

The almost inevitable outcome of the child's encounter with the quantitative environment is the construction of an elementary form of mathematical knowledge that we call "informal" (partly because it is not expressed in formal terms like written notation, and partly because it is typically not acquired through a process of formal instruction). Vygotsky (1978) called this form of knowledge "spontaneous" and declared that "children have their own preschool arithmetic, which only myopic psychologists could ignore" (p. 84). Differing from the adult's knowledge, the child's informal mathematics may take distinctive forms. At the same time, it is relatively powerful and can serve as the foundation for later school learning.

One example of informal knowledge involves adding. As early as 2 or 3 years of age, children begin to develop intuitive ideas about adding. They know that adding to a set changes it, makes it larger (Gelman, 1980). By the age of 4 or so, children begin to calculate sums involving concrete objects. For example, one study (Ginsburg & Russell, 1981) presented 4-year-old children with a task in which three objects displayed in one group were to be added to four objects shown in another. Children at this age generally calculate accurately using the strategy "counting all." This involves counting the members of both sets from the beginning. Even though already knowing from previous counting that one set contains four objects and the other three, the child counts: "One, two, three, four . . . five, six, seven."

As time goes on, children's approach to calculation evolves and matures. They spontaneously develop more efficient approaches to calculation (Groen & Resnick, 1977), eventually abandoning "counting all" for more advanced approaches, like "counting on," usually from the larger number. It is as if they get bored with counting all and discover that it is easier simply to count on from the larger number. Thus, the child begins with four, counting "four . . . five, six, seven." And of course, strategies for counting eventually get internalized, so that the child adds (by counting) solely on a mental level, not requiring concrete objects at all.

Informal notions of mathematics proliferate before entrance to school. The child typically develops notions of more and less, adding and taking away, shape and size, relative frequency, and much more (Ginsburg, 1989). Indeed, some young children even begin to think about infinity (Gelman, 1980).

Informal mathematics continues to develop throughout the life span and is often tied to context. Thus, in selling goods on the street, largely unschooled children in Brazil develop sophisticated and effective means for mental calculations; they can figure out mentally what goods should cost, make change, and eventually end up with a profit. At the same time, these children cannot perform the kinds of calculation required in schools (Carraher, Carraher, & Schliemann, 1985; Nunes et al., 1993). Research of this type has led many theorists in cognitive development to "challenge the dominant view in cognitive science that assumes a cognitive core can be found that is independent of context and intention. Instead, they argue, every cognitive act must be viewed as a specific response to a specific set of circumstances" (Resnick, Levine, & Teasley, 1991, p. 4). The thesis of "situated cognition" has relevance to theories of learning disabilities, which share with cognitive science the belief in a (defective) cognitive core.

Motivation

Children develop an informal mathematics because they find it useful (practical utility) or because they are curious about the world (intrinsic motivation). The child wants *more* food, rather than *less*; therefore, figuring out what makes more and how to get it is useful. The child observes that numbers get bigger as counting proceeds and is curious about what happens when he or she counts as high as possible. No one instructs the preschool child in adding or infinity. No one forces the child to develop the concepts and strategies of informal mathematics. Indeed, parents (not to mention psychologists, "myopic" or otherwise) are often unaware of the scope and power (and sometimes the very existence) of young children's informal mathematics.

Children usually appear to enjoy learning and to engage in it with enthusiasm. Needless to say, the effect associated with the learning of informal mathematics bears little resemblance to the discomfort many children exhibit in school.

The Widespread Nature of Informal Knowledge

Cross-cultural research shows that informal mathematics is virtually universal. The general finding is that although they differ in many important ways, children from various cultures, literate and preliterate, rich and poor, of various racial and ethnic backgrounds, all display a similar development of various aspects of informal mathematics (Klein & Starkey, 1988). Thus, unschooled African children display strategies of addition similar to those of American children (Ginsburg et al., 1981). Within the United States there are few racial or social class differences in the development of basic aspects of preschool children's concrete addition and other key informal abilities (Ginsburg & Russell, 1981).

Does this research show that all children are identical in their mathematical thinking? No. It shows instead that the general course of development of informal mathematical abilities is similar across boundaries of culture, class, and race. All children seem to develop basic mathematical abilities of the type described. At the same time, many group differences can readily be observed. For example, within the United States, poor children perform at a somewhat lower level on certain informal mathematical tasks than do their more affluent peers (Ginsburg & Russell, 1981; Saxe, Guberman, & Gearhart, 1987); but both of these groups display the same basic pattern of development.

Why do almost all children acquire the fundamentals of informal mathematics? Because, as we have seen, biology prepares them to learn, and because they live in quantitatively rich environments. Children are sense-makers, and they enjoy making sense—or at least they profit from doing so.

FORMAL MATHEMATICS

In most societies, children, already equipped with functioning mathematical minds, enter school somewhere around the age of 5 or 6. (In some societies, a form of schooling starts much earlier.) Schools are artificially created social institutions designed to pass on to children the accumulated social wisdom, one aspect of which is formal mathematics. In contrast to children's informal system, formal mathematics is a written, codified body of material conventionally defined and agreed upon. Formal mathematics is a "scientific system"—coherent, explicit, organized, and logical. We adults want children to learn formal mathematics because of the intellectual power it can grant (not to mention its beauty). Formal mathematics contrasts sharply with children's informal, "spontaneous" system—intuitive, emotional, implicit, and tied to everyday life (Vygotsky, 1986).

But the course of learning formal mathematics is often far from smooth. Here the story is about how children, already possessing an informal mathematics, enter a new and specially designed environment, the artificial culture of the school, and there encounter the world of academic mathematics. I will begin by describing the ecology of schooling, and then turn to an account of what children learn in the academic setting.

The Ecology of Schooling

Children's learning of mathematics cannot be understood in isolation from the larger ecology—the culture, schools, teachers, and texts that constitute U.S. education (see Note 2).

Culture. The dominant orientation of U.S. culture can fairly be called "math-phobic." People are generally uncomfortable with mathematics (indeed, fearful of it). At a parent–teacher meeting at my child's school, I observed rather well-educated, affluent parents whispering to each other that they could never abide math and would be unable to understand what the teacher was about to demonstrate—at the third-grade level! Visitors to a household will let a child know, albeit in a joking manner, that math homework is akin to a form of torture.

Stereotypes abound: Math is for nerds; football is for real men. Girls: Forget it. African Americans and Latinos cannot succeed in mathematics or technical professions.

Members of some other cultures value mathematics, and education generally, far more highly than do many Americans. In Japan, for example, partly as a result of the Confucian tradition, teachers are highly respected and paid rather well. And it is expected that all Japanese can and should learn mathematics, as they generally do.

Schools. In the United States, many schools are decrepit, overcrowded, poorly supplied, and, in general, terrible places in which to spend any amount of time, let alone attempt to learn a subject as demanding as mathematics. These conditions are disproportionately characteristic of schools in low income areas. Education in the United States is characterized by "savage inequalities" in the educational opportunities it affords its children

(Kozol, 1991). Poor and minority children are more likely than their White, middle class peers to attend inadequately equipped—indeed truly atrocious—schools. Furthermore, in the 1990s, the standard of living is generally eroding, with the result that cuts are being implemented in education budgets across the country, even in middle class areas.

Teachers. At the elementary-school level in particular, many teachers are uncomfortable with mathematics and share the general U.S. aversion to it. They are likely to spend less time at teaching mathematics than reading. But even worse, teachers often do a bad job at teaching mathematics. Research shows that many have a poor understanding of the subject matter and that even the surprisingly small proportion of teachers who do understand the material may not explain it well to children.

> Our results indicate that a multilevel problem exits. The first and primary one is the fact that many teachers simply do not know enough mathematics. The second is that only a minority of those teachers who are able to solve these problems correctly are able to explain their solutions in a pedagogically acceptable manner. (Post, Harel, Behr, & Lesh, 1991, pp. 195–196; see Note 3)

Many teachers think of mathematics learning as a rote activity, an occasion for frequent drill (Thompson, 1992). Asian teachers tend to explain mathematics subject matter in more depth and place a greater emphasis on thinking and understanding than do American teachers (Stigler & Perry, 1988).

Texts. Despite the forward-looking recommendations of the National Council of Teachers of Mathematics (1989), most teachers continue to employ traditional textbooks, parts of which are confusing and do not make a great deal of sense. I cannot back up this assertion with "scientific evidence," but I suggest that you examine some of the texts with an eye toward determining whether the explanations are adequate, let alone clear.

The Central Problem of Mathematics Education

The environmental conditions I have described are hardly auspicious for the learning of mathematics. Indeed, one might say that U.S. children are educationally at risk—at the mercy of (a) a culture that devalues mathematics, (b) inhospitable schools, (c) teachers who teach badly, and (d) textbooks that often make little sense. And of course, the risk is greater if the child is poor or from an underprivileged minority.

But even under favorable conditions, the task of teaching mathematics is by no means simple. Children arrive in the classroom with an informal mathematics, a set of intuitive ideas that are useful and accurate in many ways but require elaboration. It is the teacher's responsibility to help children advance beyond their initial informal mathematics. The teacher cannot leave the learning of mathematics entirely in children's hands (or minds) but must instead intervene so as to lead children to "reinvent" formal mathematics—to construct ideas and procedures that would not have arisen spontaneously in children's minds in the absence of adult help. After all, what are teachers for?

But how to do this? How can we respect children's constructions but help the children to progress beyond them? That, I think, is the basic question of mathematics education. Although the question is too immense to deal with here, it seems clear that one particular approach usually does not work: The traditional approach is to teach mathematics by defining it purely in terms of the formal system of written mathematics—by introducing explicit definitions, theorems, and the like. This may be how mathematicians eventually come to formulate mathematics, but it is not how they usually arrive at such notions, and it is certainly not a good way to teach children. Except in the minds of the most advanced students, definitions in formal terms are usually mere words with little meaning. (As one child put it after failing to solve a problem describing a "real life" situation: "I didn't know you meant for real. I thought you meant for math.") Yet this is probably the way most children are taught, often with disastrous results.

What Children Learn

Under the conditions just described as prevalent in U.S. education, ordinary children are at serious risk of failure in mathematics learning. Consider, first, achievement test data.

Achievement Tests. According to recent international comparisons, children from various Asian countries exhibit mathematics performance superior to that of children from the United States. For example, within the first few grades of elementary school, Chinese and Japanese children outperform Americans on standard tests of mathematics achievement (Stevenson & Stigler, 1992). Indeed, Japanese children outperform U.S. children even at the kindergarten level; however, Chinese (from Taiwan) and U.S. kindergarten performance is about the same (Stevenson, Lee, & Stigler, 1986).

Group differences other than those involving nationality are equally dramatic. School achievement varies by race and culture within the United

States. In general, African American and Hispanic children eventually do poorly in school mathematics and science (Natriello, McDill, & Pallas, 1990), Asian American children do extremely well (Arbeiter, 1984), and White children fall in between the two extremes (National Center for Education Statistics, 1990).

Another pervasive achievement difference involves social class. In the United States, poor children, as a group, generally perform less adequately in school than do more affluent children (Natriello et al., 1990). In this regard, Asian Americans are something of an anomaly: Their relative lack of income does not seem to detract from their school performance.

Looking More Closely. Achievement test scores provide only a general and crude view of mathematics learning. More detailed studies, often involving the "cognitive clinical interview" method, offer a richer account of children's mathematical thinking. Such research illuminates five key areas of mathematics learning:

1. Difficulty with symbolism. Many children do not understand what mathematical symbols refer to. For example, I asked Toby, a 6-year-old first grader, about the meaning of the mathematical symbols + and = (Ginsburg, in press).

Interviewer (I): OK. Can you tell me, like in this one right over here, we have three plus four equals seven, what does plus mean?

T: I'm not sure.

I: Huh?

T: I don't know.

I: I mean, does plus tell you to do something . . . what is it all about? *[A pause; she shrugs]* Not sure? What about equals? What does equals mean?

T: It tells you, um, I'm not sure about this . . . .

I: Uh-huh . . . .

T: I think . . . it tells you three plus four, three plus four, so it's telling you, that, um, I think, the, um, the end is coming up . . . the end.

I: The end is coming up . . . . What do you mean, the end is coming up?

T: Like, if you have equals, and so you have seven, then. *[She is gesturing to the problem on the table]* So if you do three plus four equals seven, that would be right.

I: That would be right, so equal means something is coming up . . . like the answer. *[We both laugh]*

So Toby cannot explain the meaning of the plus sign, although it was evident that she in fact added in response to it. Further, she states that an equals sign means "the end is coming up." If + shouts, "Add up those numbers," = screams, "Put the answer here." Toby is not unique in her approach. Most children say that = means "makes" (as in 5 + 3 *makes* 8) or that it means "Get the answer."

From the teacher's point of view, the child should know what + *means,* and also should not believe that = means "The end is coming up." The child is supposed to learn in school that = refers to *equivalence,* to the fact that the expression on one side of the equals sign is equivalent in value to that on the other.

This difficulty, of course, is not limited to Toby: Many children have problems in learning the meaning of mathematical symbols (Ginsburg, 1989; Hughes, 1986).

2. Bugs. Several studies show that children's errors in computation may result from systematic strategies, sometimes called "bugs" (an analogy with flawed software). One of the clearest examples of such a bug is this:

$$\begin{array}{r} 12 \\ \underline{-4} \\ 12 \end{array}$$

How can you subtract 4 from 12 and get 12? Almost any first grader knows that if you start with 12 eggs and take 4 away you do not end up with 12! A child can get an answer like this because he or she is not trying to solve a real problem about removing eggs from a carton. Instead, the child is solving a *math* problem, and is using the buggy strategy "Always subtract the smaller number from the larger one." The child believes that because one cannot take a larger number away from a smaller, 2 must be subtracted from 4, which of course gives 2. Then take away nothing from 1, which of course gives 1, so the answer must be 12.

What could be more logical? You could say that the child is making sense out of a new subtraction problem by assimilating it into what he or she already knows about math. Indeed, the teacher may have *said,* "Always subtract the smaller from the larger." The child's "knowledge" may well stem from a misinterpretation of the teacher's remarks, but the bug results from an attempt to make sense of new problems, given the (faulty) knowledge available.

Unfortunately many teachers do not look below the surface of wrong answers and simply accept them as such. Many teachers think that a wrong answer is simply a wrong answer. They do not realize that the child's wrong answer can be an intelligent answer, a sense-making effort. Many teachers take wrong answers to indicate a lack of knowledge of mathematics, or even stupidity. But that may not be the case: Wrong answers of this type may well indicate the presence of mathematical knowledge, an effort at intelligent response. Of course, not all wrong answers are "intelligent" bugs like these. Some wrong answers result from wild guesses, and others from mere sloppiness. But some errors clearly result from systematic strategies and principles.

3. *Beliefs.* Children often develop harmful beliefs about what is entailed in the learning of mathematics (Baroody, 1987; Schoenfeld, 1989). Consider again an example provided by Toby (Ginsburg, in press). She told me that in class she was using a "robot book," which seemed to be some kind of workbook. Although the details are of little interest, Toby's general views of schooling are worth noting.

I: I don't understand what the robot is . . . .

T: OK, it's like a book . . . .

I: Yeah.

T: . . . and it has math in it to help you with it. *[She showed how she solved a robot book problem]* . . . See that's how you do it . . . and if I made a mistake she would do this *[She made a sad face]*, and if I didn't, if I was all correct, she would do this *[She made a happy face]*.

Clearly the teacher focused on the right and wrong answers. I then asked her why the problem was written in a certain way.

T: They do it different ways . . . they do it any way they want.

I then suggested that the problem might be written in that form for a certain sensible reason.

T: It . . . no, they try to make it, um, tricky.

I: Tricky . . . they try to trick you? *[She nods]* How do you like that? *[She smiles]* Why do they try to trick you?

T: Because they want to make sure you know your math very well.

From experience in the classroom, Toby has acquired a view of mathematics education that might be paraphrased as follows. To help the child learn mathematics or to assess his or her knowledge of mathematics ("they want to make sure you know your math very well"), the teacher (or the textbook) presents problems that are both arbitrary ("they do it any way they want") and deceptive ("they try to make it, um, tricky"). From the child's point of view, the teacher's role is to create obstacles, to present meaningless tasks, to trick, and then to reward the child's success with praise or punish failure with disapproval.

Other children hold equally harmful views of mathematics learning. Many see it as a rote activity (Schoenfeld, 1989), and, probably from experience with flash cards, some believe that math is that subject in which one is supposed to get correct answers quickly without thinking (which is seen as cheating).

4. Rote learning. It should not take a great deal of psychological research to convince us that a good deal of mathematics learning is rote. The child learns procedures that succeed in solving problems but seldom understands the rationale for the procedure. Here is a simple example involving Chris, in Grade 2. He had just written that 9 + 5 = 14.

I: Now I'm going to ask you something about 14. How come you wrote 14 with a 1 and then a 4?

C: 'Cause that's how I write 14.

I: I notice that when you write 14 you have a 1, and on the right of that is a 4. What does that 4 stand for?

C: 'Cause it's 14.

I: All right. What does the 1 stand for?

C: That's how you write 14.

I: Why don't you write it like this [41]?

C: That's 41.

I: All right. Why do you write 41 like that?

C: Because there's a 4 there and a 1 there.

I: Why did they invent that way of doing it? What could it possibly mean? What does that 1 stand for?

C: 1.

I: What does that 4 stand for?

C: 4.

I: Can you write the number one hundred twenty-three? That's right. What does that 1 mean?

C: 1.

I: Just 1. And what does that 2 mean? What does it stand for? What is it telling us?

C: 2.

I: Just 2. And the 3? What does that tell us?

C: Just 3.

Chris has the rules necessary for the accurate writing of numbers but can say next to nothing about the ideas behind them. His work with numbers seems entirely on an action level and does not appear to encompass any higher levels of understanding. Chris can *operate* but not *theorize*. This type of phenomenon occurs at all levels of mathematics, from the simple writing of numbers to the operations of algebra and beyond.

5. *Lack of connection between the informal and the formal.* Many children have some informal mathematical knowledge but fail to use it in dealing with written mathematics problems. Here is a simple example involving a fifth grader, Bob.

I: Let's see, now, 21 take away 5.
  Bob wrote:

$$\begin{array}{r} 21 \\ \underline{-5} \\ 24 \end{array}$$

I: Do you think that's the right answer?
B: *[He checks his work]* Yes.
I: If you had 21 candies and gave away 5, how many would you have left?
B: *[He quickly solved the problem in his head]* Oh! I think I added. *[He seemed to mean that since he got more than 21 he must have added]*
I: Do you still think that's the right answer?
B: No. 'Cause 5 take away from 21 is 16.
I: *[The interviewer pointed to his written 21 - 5 = 24]* But why isn't that the right answer?
B: I don't know.
I: You had 21 and you took away 5, and you had 24 left. What's wrong with that?
B: Oh! I know! It's from this one *[the 1 in 21]* that it was supposed to be taken away from.

In brief, Bob first used an incorrect written procedure for subtraction. This gave him 21 - 5 = 24, where the result is larger than the number he started with. He easily did the same problem in his head by counting backwards and in this way got the right answer. Bob then realized that his written answer was wrong: "Oh! I think I added." This shows that he knew that in subtraction he should end up with less, not more, than he started with. Then he saw why he was wrong in the first place: He should have subtracted the 5 from the 1, rather than the 1 from the 5.

Like many other children, Bob was living in two worlds. One was Vygotsky's (1986) world of spontaneous mathematics, a world in which he could get correct and sensible answers by counting backwards mentally; the other was Vygotsky's world of scientific mathematics, a world in which the child used buggy procedures like taking away the smaller number from the larger in order to get incorrect and absurd results. Like many other children, Bob usually did not make connections between the two worlds. He could do his sensible, informal math largely in the everyday world; and he saved his absurd, formal math for school. The educational system was not successful in helping him to connect the two worlds.

Summary

We have seen that ordinary children are at risk in the U.S. educational system. Although possessing an adequate informal mathematics, they often encounter a culture that tends to devalue mathematics, schools that are inadequately equipped, teachers who do not like or understand mathematics, and textbooks that make little sense. The results are that U.S. children perform poorly relative to children in other countries, and that within the United States, ethnic and social class differences are large. Furthermore, a closer examination of thought processes reveals that ordinary children (a) experience difficulty in using symbolism, (b) use buggy strategies, (c) hold harmful beliefs about mathematics, (d) engage in rote learning, and (e) fail to connect their informal mathematics with what is taught in school. Children's difficulties tend to cumulate over the years. As mathematics becomes more complex, children experience increasing amounts of failure, become increasingly confused, and lose whatever interest and motivation they started out with.

MATHEMATICS LEARNING DISABILITIES

To this point, I have described the context within which children typically study mathematics in the United States, and I have outlined some achievement and cognitive outcomes. All this provides the background for considering the central topic of this chapter, namely, issues of learning disabilities in mathematics. I begin by describing the goals of research on learning disabilities and then show how the classical approach is not adequate for achieving them.

Goals

The researcher's goal is to understand a special subset of children—those who seem to have unusually severe problems in learning mathematics. These children with "learning disabilities" possess "normal" intellectual ability—they are not mentally retarded—but do not seem to profit from sound instruction, despite the fact that they are motivated to learn (although, after a period of failure, they may lose whatever motivation they began with). These children seem to be trying hard, but they fail, presumably at least partly as a result of some kind of cognitive difficulty. Of course, it makes no sense to suppose that merely labeling these children as learning disabled explains anything: In general, the use of diagnostic labels, definitions, and categories (learning disability, schizophrenia, etc.) is only the first step in understanding a phenomenon—or, more precisely, it identifies a phenomenon that needs to be understood. The goal then is to understand the cognitive processes and other factors underlying the academic failure of children with learning disabilities.

This goal presents the researcher with two challenges. One is to identify the subset of children whose failure needs to be explained. Children with genuine learning disabilities must be distinguished from those who display garden-variety poor mathematics performance. It is necessary to articulate a concept of learning disability that allows us to identify mathematics learning difficulties that are not caused by the ordinary shortcomings of mathematics instruction, but are instead related to specific cognitive factors. The second challenge is to provide a useful conceptualization of the cognitive difficulties experienced by children with learning disabilities. How are their cognitive processes different from those of nondisabled children who do poorly in mathematics?

The Traditional Approach

The traditional approach to meeting each of these challenges suffers from basic weaknesses. These include mislabeling, the defective concept of "defect," and the weakness of neurological explanations.

Identifying Children with Learning Disabilities. The traditional approach to identifying the population of interest is to select children of normal IQ who receive low achievement scores in ordinary schools. In practice, the researcher eliminates from consideration children classified as emotionally disturbed (ED); everyone else is fair game for the learning disabilities label. This results in frequent diagnoses of learning disabilities, in both reading and mathematics. Approximately 6% of elementary school and junior high school children are diagnosed with a mathematics disorder, as compared with approximately 5% with a reading disability (Badian, 1983; Geary, 1993).

The traditional approach suffers from two major flaws. One is that it assumes that ordinary schools provide adequate instruction. For example, the World Federation of Neurology proposed that dyslexia "is manifested by difficulty in learning to read despite *conventional instruction* [italics added], adequate intelligence, and sociocultural opportunity" (Shaywitz, Escobar, Shaywitz, Fletcher, & Makuch, 1992, p. 145). But "conventional instruction" is precisely the problem: Ordinary mathematics instruction is simply not very good in the United States. The general conditions of mathematics education are so deplorable that many (if not most) children of normal intelligence experience significant difficulty in learning mathematics. The most reasonable explanation for a nondisabled child's failure in mathematics is the conventional instructional system—the textbooks, teachers, cultural atmosphere, and curriculum.

A second flaw is that the system of selecting children on the basis of IQ and achievement results is overly broad: It does not exclude enough children. Even under conditions of adequate instruction, children with normal IQs may fail in school for many reasons other than cognitive deficit (e.g., they may be poorly motivated, or they may have a poor self-concept as a learner).

These two flaws result in the misidentification of large numbers of children as learning disabled. Indeed, one authority (Farnham-Diggory, 1992) suggested that the mislabeling of children as learning disabled is rampant in the United States: "By recent estimates, 80 percent of the children who are classified as learning-disabled should not have been" (pp. 5–6). This estimate may or may not be high, but common experience (mine at least) suggests that misclassification is not uncommon (see Note 4). If this is the case, then the classical approach has the effect of muddying the research waters: How can researchers produce consistent and sensible results if their samples include many children who are not in fact learning disabled?

In brief, the classical approach to identifying children with learning disabilities is faulty in several respects. It assumes that ordinary instruction is "adequate," and that a discrepancy between achievement and IQ (given "adequate" instruction) is a sufficient basis for selecting children with learning disabilities. This has the effect of misidentifying many children, particularly the many children whose poor performance can be explained simply on the basis of the inadequate mathematics instruction prevalent in U.S. schools. The result is that researchers' samples of children with learning disabilities are "impure." Many of the participants are in fact not learning disabled at all.

The Concept of Defect. The traditional view goes on to assume that the basic cause of poor performance in children with learning disabilities is a cognitive defect. If the child's intelligence is normal, if instruction is adequate, and if "sociocultural opportunity" is present, then any failure to learn must result from some defect within the child, namely, a specific incapacity, a cognitive disorder. The traditional view sees this defect as residing in the child—a characteristic that is the immediate cause of poor mathematics performance in children of normal intelligence. In this "reductionist" scenario (Poplin, 1988), the defect then needs to be fixed.

The problem with this point of view is that learning disabilities cannot be understood in the overly simple terms of a defect or set of defects in need of repair. Cognition is "situated." We cannot consider it apart from the many contexts in which it operates. To understand the child's failure, we need to consider how his or her cognition functions in its context, particularly the context of schooling, which, as we have seen, cannot be assumed to be adequate.

The Role of Neurological Explanations. A fruitful neurological explanation of learning disabilities must involve at least two components. One is an account of the particular defective cognitive processes that are the immediate causes of poor performance in children with learning disabilities. It is first necessary to identify such processes before one can provide a neurological explanation for them. A second component is an empirically based account that connects the identified cognitive process to a specific neurological disorder or area of brain damage. At present, neither of these components is readily available. Research (to be reviewed below) has provided little information concerning the cognitive processes underlying mathematics learning disabilities, and direct evidence linking defective cognitive processes with specific neurological disorders or areas of brain damage is minimal (although recent advances in neurological measurement promise progress in the near future). Hence, at the present time, we lack useful neurological theories for explaining mathematics learning disabilities.

A Developmental Perspective

Consider first how a developmental perspective contributes to the identification of children with learning disabilities, and then to the conceptualization of their difficulties.

Identifying Children with Learning Disabilities. The goal at the outset is to identify a specific group of children—those with "real" learning disabilities—whose behavior needs explaining. Doing this requires eliminating from consideration those children whose failure can be explained on the basis of inadequate instruction, poor motivation, and other noncognitive factors.

The first issue, then, is whether the children under consideration are in fact receiving adequate instruction. To find out, it is necessary to examine the process of mathematics instruction in their classrooms. If that instruction is particularly inappropriate or inadequate, children's poor achievement need not be attributed to learning disabilities. Is it easy to evaluate instruction in the classroom? No. And at present it is not even practical to do so on a systematic basis. But because understanding children's experience of education is vitally important, researchers need to develop systematic methods for analyzing and evaluating classroom instruction.

A second step is to examine closely, within the "laboratory" setting, certain possible causes of the child's poor achievement. Suppose a child is presumed to have difficulty in learning simple addition. Does the failure to learn result from lack of attention, inadequate motivation, or some other cause? If these factors can be ruled out, then does this child indeed exhibit significant difficulty in mastering simple addition when given a reasonable period of adequate instruction? If so, then the researcher can proceed with further efforts at understanding the cognitive source of the difficulty.

An alternative to the laboratory investigation of the child's ability to profit from instruction was suggested by Geary's (1990) research study. Geary began by selecting first- and second-grade children defined as learning disabled (LD) by their schools. These children were receiving remedial education in mathematics for about 20 minutes a day. Then the investigator used the current year's achievement test scores to separate these children into two groups: LD–improved and LD–no change. The former group's scores had improved to the point where the children placed out of remedial education; the latter group's scores were substantially stable. Further, the results showed that the LD–improved group displayed underlying cognitive processes similar to those of nondisabled children. The LD–improved group were perhaps "developmentally delayed" but not "developmentally different," as were members of the LD–no change group. The investigator concluded that the "initial poor achievement scores of the LD–improved group were likely due to inadequate preacademic skills . . . and/or the initial misclassification of some of these subjects . . . and not due to an underlying cognitive or meta-cognitive deficit" (p. 378). One might say that the LD–improved group were not really learning disabled. Geary's design is well worth considering in the effort to eliminate false positives in future studies of learning disabilities.

In brief, the first goal is to identify children with "real" learning disabilities. Doing so might involve observation of teaching practices in their classrooms, close examination of their behavior in laboratory learning situations, and consideration of their response to remedial teaching in the school setting. Clearly, the researcher cannot accurately identify these children simply by noting a discrepancy between IQ and achievement and by assuming adequate instruction.

Suppose that the learning disability seems genuine. The next step is to examine the many different cognitive factors that may be involved.

Many Processes Underlying Different Learning Disabilities. Mathematics presents many faces, for example, arithmetic is vastly different from topology, which in turn has little in common with probability. Each of these very different topics may require different mental processes. Mathematical thinking is not unitary. Learning to count involves one set of cognitive skills (i.e., initially memorizing the numbers 1 to 12, and then detecting rules underlying the language), whereas appreciating the notion of equivalence requires other cognitive operations (perhaps something like Piaget's [1952a] concrete operational "reversibility"). Even within a single type of problem, children display a variety of problem-solving strategies (Siegler, 1988).

This complexity makes it likely that there is no *single* cognitive defect that causes failure in learning mathematics. There are probably many different learning disabilities in the sense that various cognitive processes may be implicated in learning difficulties with respect to different aspects of school mathematics. Presumably, the difficulties in learning to count are different from those encountered in learning basic principles of equivalence in algebra. It is therefore necessary for the researcher to investigate the individual child's understanding of the various mathematical topics he or she is attempting to master—writing numbers, concepts of addition and subtraction, calculation, number facts, and the like.

How to do this? The developmental literature provides many leads. First of all, it is useful to consider the child's informal knowledge in each of these areas. Does the child have available adequate *informal* notions of adding, quite apart from the written calculations learned in school? As I have pointed out, research shows that informal mathematics is quite widespread among normally achieving children. Is the same true of children with learning disabilities? Second, do the child's errors result from ordinary bugs, unusual error strategies, or the lack of strategy? Are the solution strategies of children with learning disabilities unusual or merely immature (Geary, 1990)? Third, do memory difficulties play a large role in the child's failure? One of the most consistent research findings is that children with LD have a particularly difficult time remembering basic number facts (Geary, 1993; Russell & Ginsburg, 1984). In fact, that deficit seems to be the major feature differentiating children with and without learning disabilities. Fourth, whether or not they compute accurately, do children with learning disabilities understand basic mathematical concepts (e.g., that adding makes sets greater in number)? Do children with learning disabilities make connections between their informal ideas and the written procedures and symbols essential to school mathematics?

An interesting problem for further research is to compare children's difficulties across different aspects of mathematics. Suppose the child displays unusual (not simply immature) strategies in written addition. Is he or she likely to exhibit similar strategies in written subtraction? Or, if basic informal concepts are lacking in addition, will they also be lacking in geometry? Mathematics learning disabilities may or may not be consistent across different topics.

Sensitive Methods. Suppose, then, that the researcher wishes to examine different aspects of the child's mathematical knowledge in the ways described above. What methods should be used to accomplish this? In developmental psychology, investigators are increasingly likely to employ a variety of methods designed to investigate the child's thinking with greater sensitivity than is afforded by traditional standardized tests. For example, researchers are currently attempting to develop new testing procedures that can measure underlying understanding, strategies, and the like (Ginsburg, 1990; Glaser, 1981; Royer, Cisero, & Carlo, 1993). Other methodological innovations are more far-reaching. For example, researchers have gone outside of the laboratory to investigate memories of everyday life (Pillemer & White, 1989); they have employed talk-aloud methods to examine complex problem solving (Ericsson & Simon, 1993; Schoenfeld, 1985). The clinical interview method—that is, the researcher's deliberately nonstandardized and flexible questioning of the child—has emerged as a major research tool in the investigation of mathematical thinking (Ginsburg, Jacobs, & Lopez, 1993). The method of "microgenesis" (repeated observations and interviews of individual children working on a set of problems over a relatively long period of time) has been used to examine development (Kuhn & Phelps, 1982; Siegler & Crowley, 1991). Others have called for methods that will allow for consideration of meaningful activities in social context (Kuhn, 1992). Ethnography—the detailed observation of persons in their natural environments and cultures—is traditional in anthropology and has much to offer research in some areas of psychology (Fetterman, 1989).

Innovations like these have opened up developmental research methods, freeing them from traditional constraints. The field of learning disabilities research can also profit from innovations like these. It is not necessary for researchers to be limited to what used to be considered "scientific," namely, standard tests of one sort or another. Such instruments often fail to reveal much that is interesting about children's thinking.

Teaching Experiments. One method of particular relevance to research on learning disabilities is the *teaching experiment.* Suppose that we have discovered that children with learning disabilities fail number-fact problems because of a memory difficulty. This is important information, but it is not enough. The next logical question might refer to the seriousness or depth of the memory difficulty. Is it stable and relatively permanent, or can it be easily overcome? If the cognitive disorder can be removed with a minimum of effort, it was not very fundamental in the first place, and this information is important for the evaluation of the child's learning disability.

The most direct way to investigate stability is through some form of teaching experiment, or what developmental psychologists refer to as the "zone of proximal development" (Vygotsky, 1978), defined as "the distance between the actual developmental level, as determined by independent problem-solving, and the level of potential development, as determined through problem-solving under adult guidance, or in collaboration with more capable peers" (p. 86).

In the United States, several investigators have expanded upon Vygotsky's (1978) theory of social learning. In one line of work (Campione, Brown, Ferrara, & Bryant, 1984), an analysis of students' mistakes on a pretest suggests to the examiner the areas in which learning potential can be assessed. Following this, the examiner presents the student with a problem in an area in which he or she has shown difficulty and provides hints to assist in the problem-solving process. These range from general metacognitive hints to those specific to the demands of the task. The amount of help the student needs is an estimate of learning efficiency within that domain. The examiner continues presenting the student with problems similar in nature to the initial one, providing as much help as needed, until the student is able to solve problems independently. After helping students to achieve independent learning, the examiner presents near, far, and very far transfer problems and students are given assistance as needed to solve them. In some cases the students are given a posttest to assess the gains following the assessment. Techniques like these are essential to investigating the extent to which cognitive difficulties persist despite persistent efforts to remove them.

The Role of Context. Suppose that various assessment techniques, including teaching experiments, document that a child has an enduring difficulty in some aspect of mathematical cognition. To understand the difficulty even more fully, researchers must consider the role of context. As we have seen, recent developmental theory stresses that cognitive development always takes place within a social and cultural setting (Resnick et al., 1991). Children may show sophisticated addition skills in the context of bowling but not in the context of monetary transactions; they may excel at counting some kinds of objects but not others. Similarly, learning disabilities may thrive in certain situations but not others. A child may have great difficulty remembering number facts or solving word problems under certain conditions but not others. Perhaps learning disabilities are less general—and more context dependent—than we ordinarily suppose; that is an important issue for future research.

The Development of Children and of Mathematics. According to one study of reading (Shaywitz et al., 1992), "the diagnosis of dyslexia is not constant over time but will show a predictable year to year variability . . . . Only 28 percent of the children classified as dyslexic in grade 1 were also classified as dyslexic in grade 3" (p. 149). At least one kind of learning disability—dyslexia—is not necessarily permanent.

There are reasons for expecting that a similar lack of stability should characterize mathematics learning disabilities. The mathematics taught each year changes; and, as children develop, their cognitive abilities change. As mathematics "develops," it requires different types of thinking. For example, the transition from mental addition to the written symbolism of + and = presents 6-year-olds with one set of cognitive challenges; the transition from whole numbers (simple arithmetic) to the rationals (particularly notions of ratio) presents 10-year-olds with another set of difficulties.

If new topics demanding different skills are introduced as school proceeds, children with specific cognitive defects might have difficulty with some topics but not others. Perhaps children can "outgrow" some mathematics learning disabilities, and "grow into" others.

So a cognitive "defect" (e.g., difficulty in remembering number facts) is not absolute but may be relative to setting, topic, and age. A "defect" may prevent a 6-year-old from memorizing number facts presented on flash cards in the first grade but may not interfere with his or her use of money or with the geometry also taught in the first grade. And later, in the fourth grade, the defect may disappear, or if it continues, it may be irrelevant to the newly introduced ideas of ratio. Researchers need to examine not only children's performance across various mathematical tasks but also changes in children's difficulties as new topics are introduced at each grade level. Such longitudinal research is almost nonexistent in the field of learning disabilities.

The Whole Child. Learning disabilities cannot simply be *reduced* to a few discrete mental defects; they should be considered from a holistic point of view (Poplin, 1988). A functioning child's learning disability is not only the cognitive defect but also the child's feelings about it, and those feelings are in turn influenced by teachers' and parents' beliefs about the learning disability. The child does not operate in isolation; the learning disability lives in a school and in a family, whose members in turn have to live with the disability. Again, a learning disability must be considered in context (here in the context of the whole child, who in turn must be understood in the larger ecology of the family and school).

All of this produces enormous complications for research. It is very difficult to design studies that can effectively isolate the effects of particular cognitive processes. The child fails not only because of a memory defect but because the teacher reacts to it in certain ways, because the parents hold certain beliefs about the child and the defect, and because the child "constructs" his or her own concept of what it means to "have" such a defect (and to be treated in certain ways by teachers and parents).

How to deal with such complexity? One approach is to broaden research to encompass issues of belief, motivation, and the "cultural construction" of learning disabilities (McDermott, 1993). We need to consider issues like teacher and parent beliefs about LD. We need to examine the self-concept of children with LD, their patterns of motivation, their defenses. We need to consider the claim that a learning disability is a kind of "cultural construction" that can "acquire a child" because that is how the culture copes with perceived difficulties in learning in some settings (McDermott, 1993).

Of course, considering new variables does not resolve the fundamental complexity but in a sense only adds to it. However, one thing is clear: We need to abandon the traditional view in which deficit is seen as *cause* of poor performance, and in which independent variables are held to influence dependent variables in general, regardless of context. Psychological life is messy, and simple, context-free notions of cause and effect will not do. No one ever said research in this area would be easy.

Educational Strategies. A learning disability is not an incurable disease with no remedy. Research should investigate various methods for helping children overcome learning disabilities. One such approach involves bypassing learning disabilities. Certainly we all have learning difficulties, if not disabilities, to some degree (is anyone equally comfortable with or proficient at all aspects of learning?), and therefore the teacher's task is to help us find a way around them.

We have seen that mathematics is multifaceted, requiring many different types of thought, and that children may be characterized by difficulties in some areas but not in others. We have seen that cognitive defects are not necessarily general but may be tied to particular settings, topics, and age levels. If this is so, then one approach to math learning disabilities is to bypass them.

Consider again the well-documented inability to remember number facts. Assume for the moment that this inability cannot be corrected—that children who cannot remember number facts cannot learn to do so. This "deficit" is certainly a problem in learning the number facts the way they are usually taught, that is, by rote memory. But it is not clear that children with this kind of memory defect would be at a disadvantage if instruction were modified or if they were engaged in studying other branches of mathematics in which such memorization was not basic.

If the teacher stresses the rote memorization of number facts and insists that no other approach is acceptable, then the child with this particular disability—if it cannot be corrected—will experience serious difficulties in learning. But if the teacher instead attempts to bypass the difficulty by means of a focus on understanding and "figuring out" the number facts, and even allows the use of calculators (which are a kind of "prosthesis" or "assistive technology device" for children with memory or calculational difficulties), then the child with a memory disability is likely to experience no particular difficulty. The hypothesis is that whether a cognitive disorder has major effects on the child's learning depends to a great extent on how the teacher conducts instruction. In this sense, LD may be socially constructed. And of course, if that child is allowed to progress beyond the rather trivial topic of number facts to more interesting mathematics, he or she may do quite well indeed.

CONCLUSIONS

The development of mathematical thinking and the learning disabilities connected with it are complex, challenging, and important to investigate. New concepts of informal and formal thought and of cognition situated in context may help researchers to understand the development of learning disabilities, and innovations in method may allow researchers to study them.

A developmental perspective suggests that researchers concerned with learning disabilities abandon a deficit model and instead undertake a research agenda that involves the use of such sensitive research methods as clinical interview, ethnography, and microgenetic method to examine (a) children's informal knowledge, (b) their construction of different forms of knowledge in school, (c) the adequacy of ordinary instruction, (d) children's response to good teaching, (e) children's motivation, (f) the interaction between modes of thinking and educational context, and (g) the development of children's thinking over time.

NOTES

1. Without these similarities, there would be no possibility of scientific understanding. If everyone were truly unique and entirely different from everyone else, observers could make no general statements whatsoever, and, indeed, people could not communicate with one another. It is our shared humanity—our lack of complete uniqueness—that makes possible both science and ordinary social intercourse (including love).

2. This ecology differs dramatically in different countries. See, for example, Stevenson and Stigler (1992) for a description of schooling in Asian countries.

3. Similarly, Eisenhart et al. (1993) reported a case study of a teacher who, although majoring in mathematics for 3 years, "did not always have the content or pedagogical knowledge she needed to teach for conceptual knowledge, despite her desire to do so" (p. 24).

4. Why is so much misclassification tolerated? To some extent, the prevalence of learning disabilities is tied to the availability of funding. School districts will use whatever funding they can find to provide services for needy children. To the degree that funding is available for learning disabilities, children will tend to receive that diagnosis; to the extent that the availability of funding shifts to another disorder, the diagnoses will tag along.

One wonders whether schools overuse the diagnosis of learning disabilities for another reason entirely: If the child has a defect, a disease that prevents learning, then clearly the school is not responsible for educational failure. The strategy of blaming the victim has also been widely applied regarding the learning failure of poor children in U.S. schools (Ginsburg, 1986). Another curious cause of mislabeling is parental pressure. The great psychologist, Alfred Binet, one of the founders of intelligence testing, felt that being classified as "unable to profit . . . from the instruction given in ordinary schools . . . can never be a mark of distinction, and such as do not merit it should be spared the record . . ." (Binet & Simon, 1916, pp. 9–10). But Binet was wrong. Many parents do want their children classified as being "unable to profit . . . from the instruction given in ordinary schools." Why do parents want a diagnosis of learning disability? Sometimes it gets their children special benefits—for example, extra help, or release from the requirement that the SAT test be taken under timed conditions. Sometimes, I think, there are psychological benefits for the parents: Their child is not simply struggling with learning, not simply an average student; instead, their child has a certifiable disease, which clearly excuses him or her from achieving the excellence that the parents hope for (but which the child finds unattainable).

3. Disabilities of Arithmetic and Mathematical Reasoning: Perspectives From Neurology and Neuropsychology

BYRON P. ROURKE AND JAMES A. CONWAY

For many years, discussions of learning disabilities (LD) were mostly limited to unexpected developmental difficulties with functions such as reading and spelling. Although disorders of calculation have a fairly long history in the neurological literature, extending at least as far back as the early years of this century, work in this area was primarily concerned with acalculia as an acquired disorder resulting from brain damage incurred after a relatively normal course of early development. The study of dyscalculia as a developmental disorder, and more specifically as a subtype of learning disability, is of much more recent origin. Despite a massive literature on reading and other disabilities that appear to be linked very closely to disorders of language, there remains a relative lack of research concerning disabilities of arithmetic (Badian, 1983).

This state of affairs has resulted, in part, from the significance that arithmetic calculation had in early formulations of brain–behavior relationships. For example, many early neurological reports of patients with disordered calculation ability considered this symptom to be a manifestation of aphasia, and this perspective was eventually generalized to accounts of the relationship between developmental dyscalculia and dyslexia. From this perspective, arithmetic is a derivative skill having a basis in linguistic competencies, and the persistence of this assumption has undoubtedly hampered progress in the study of arithmetic disabilities. Moreover, social and cultural factors have influenced our evaluation of the relative importance of this topic. It has been suggested that having deficient arithmetic skills is generally considered to be more "socially acceptable" than having an impairment of reading, writing, or spelling (Cohn, 1968).

Several investigators have determined the prevalence of arithmetic disability to be at least 6% (e.g., Badian, 1983; Kosc, 1974). Kosc studied a large sample of Czechoslovakian children and found that 24 of 375 (6.4%) fifth graders were dyscalculic according to his definition. Badian reported incidence rates of poor achievement (a score at or below the 20th percentile on the Stanford Achievement Test) for a sample of 1,476 children in Grades 1 through 8, and concluded that 2.2% were low only in reading, 3.6% only in mathematics, and 2.7% in both reading and mathematics. The total number of students who demonstrated poor arithmetic ability with or without associated reading difficulty was 94 (6.4%), which is identical to the incidence rate reported by Kosc. It is clear that difficulties with arithmetic are by no means rare. Given the arithmetical demands of education, employment, and the many activities of daily living, current estimates of the prevalence of arithmetic disabilities should be taken as further evidence of the need for continued research in this area (Keller & Sutton, 1991).

In this chapter, we examine the historical and conceptual roots of arithmetic disabilities as these have unfolded in the literature of neurology and neuropsychology. We attempt to outline what is known about the neurological substrates of calculation, with an eye toward relating this information to the neuropsychological correlates of disordered arithmetic learning ability. Acquired acalculia and developmental disabilities of arithmetic and mathematical reasoning each raise some interesting questions concerning lateralization and localization of function. Furthermore, disabilities of arithmetic learning have provided particularly good examples of how the identification of specific subtypes of LD has led to a richer understanding of the unique patterns of neuropsychological assets and deficits displayed by these children, as well as to developmental models of central processing deficiencies in children (Rourke, 1982, 1987, 1988, 1989; Rourke & Fisk, 1988).

The implications of these views for research and intervention with children with arithmetic disabilities flow from two premises. First, a distinction can be drawn between calculation as a discrete function and the more general context of arithmetic/mathematics learning and performance in which it occurs. Second, mathematical reasoning is but one dimension of the more general concept-formation and problem-solving skills necessary for successful learning, and a distinction can be drawn between the brain–behavior connections pertinent to successful acquisition of arithmetic skills and those on which well-learned performance depends.

HISTORICAL OVERVIEW

Acquired Acalculia and the Neurological Approach

Notions of the brain as the organ of thought and behavior received considerable prominence in the writings of Descartes (1596–1650), who "localized" the mind in the pineal gland. Interest in the relationship between the human brain and calculation ability can be traced back to the first attempts at localization of function, specifically to the phrenological theory of Franz Josef Gall (1758–1828) and Johann Casper Spurzheim (1776–1832; see Kolb & Whishaw, 1990; Levin, Goldstein, & Spiers, 1993). The resemblance of phrenological theory to modern theories of localization of function warrants the following discussion.

The anatomists Gall and Spurzheim believed that they could ascribe particular functions to different parts of the brain by examining the various bumps and depressions of the skull and correlating them with an individual's behavioral characteristics. A convexity was assumed to reflect a well-developed underlying cortical gyrus, responsible for a particularly well-developed behavioral function, and a depression indicated underdevelopment of that area and its functional correlate. Despite the fact that the outer surface of the skull does not mirror its inner surface, much less the surface of the brain, Gall and Spurzheim proceeded to "locate" a number of behaviors, including calculation. These investigators apparently found that the temporal area of the skull, just behind and above the eye, tended to show a protrusion in mathematicians and mathematical prodigies, leading them to conclude that the organ of calculation is located "in a convolution on the most lateral portion of the external, orbital surface of the anterior lobes" (cited in Levin et al., 1993).

Phrenology was quickly dismissed by the scientific community and was replaced by more sound methodologies including the experimental ablation techniques of Pierre Flourens (1794–1867) and the clinico-anatomical correlations of Paul Broca (1824–1880). Flourens's experiments with animals argued against localization of function but nevertheless set the stage for the rapid advances that would begin with the work of Broca in the 1860s. Broca demonstrated that damage to the third frontal convolution of the left hemisphere could abolish speech, and thus began the scientific study of localization of function in the human brain.

The neurological approach to brain–behavior relationships has traditionally been greatly concerned with issues of localization, drawing upon both idiographic and nomothetic observations in order to relate specific behavioral deficits to focal lesions of the central nervous system (CNS). Many early investigators relied heavily on individual case studies, partly because comparable subjects were few and far between and statistical knowledge was limited. The detailed examination of an individual case has remained a common and fruitful method for examining disorders of calculation, having certain advantages over a contrasting-groups approach. Of particular importance to the present context are the rich descriptions of the different types of arithmetic errors made by individuals whose disabilities might otherwise be lumped together under the general heading of acalculia. Error analysis of impaired arithmetic has resulted in more sophisticated classification systems, as well as in greater understanding of the component processes involved (Spiers, 1987).

Lewandowsky and Stadelmann (cited in Levin et al., 1993) were the first to publish a detailed case study that focused on an acquired disruption of calculation ability, distinct from aphasia and resulting from focal brain damage. Their patient had a right homonymous hemianopsia (no vision in the right half of the visual field) and difficulties with both written and mental calculation. The patient was described as often being unable to recognize arithmetic symbols, despite intact ability to follow the necessary computational procedures. Based on their observations of this patient, Lewandowsky and Stadelmann suggested that a specific type of alexia for numbers could result in a person being able to recognize individual digits while being unable to read several combined digits as a single number. Levin et al. suggested that this resulted from an inability to apply the rules of the propositional system. Believing their patient's difficulties to be based in visual factors, Lewandowsky and Stadelmann proposed the left occipital region as the "centre for arithmetic faculties." Their paper was historically significant in that it was the first to propose that disorders of calculation resulted from a focal lesion that was distinct from one producing aphasic symptoms. Furthermore, they described a specific type of alexia for numbers that they considered to be separate from alexia for letters or words (Levin et al., 1993).

The first statistical analysis of a large number of cases was reported by Henschen (cited in Levin et al., 1993), who was also the first to apply the label "Akalkulia" to disturbances of computational ability associated with brain damage. He suggested that the neural substrate for calculation was distinct from, but proximal to, that of language, with lesions of the left angular gyrus being implicated in nonaphasic patients exhibiting alexia and agraphia for numbers. Henschen analyzed 305 cases of calculation disturbance reported in the literature, in addition to 67 of his own patients, and determined the existence of a small subgroup of persons in whom brain damage had resulted primarily in a disturbance of calculation, with little or no aphasic symptoms. Similar results were obtained by others (e.g., Singer & Low, 1933), providing further evidence that the neural substrate of calculation ability was anatomically distinct from that of language, and that the deficits producing acalculia may occur independently of aphasia (Badian, 1983; Levin et al., 1993).

Henschen's observations were soon followed by the work of Berger (1926), who proposed the distinction between primary and secondary acalculia. According to Berger, primary acalculia refers to a specific disruption of calculation ability and cannot be attributed to more generalized difficulties in prerequisite abilities, such as short-term memory or sustained attention. Secondary acalculia, on the other hand, refers to a symptom resulting from either a specific primary deficit (e.g., aphasia) or a more pervasive disruption of brain function. Generalized brain dysfunction may disrupt calculation performance via impairment of any number of prerequisite skills and abilities, including language, memory, attention, and cognition (Levin et al., 1993). According to Berger, primary acalculia is attributable to posterior left-hemisphere lesions not necessarily invading the angular gyrus, whereas secondary acalculia results from diverse focal lesions, or generalized damage (Benton, 1987).

In a seminal work on the classification of acalculias, Hécaen, Angelergues, and Houillier (1961) performed a detailed error analysis and proposed a tripartite organization based on the presumed neuropsychological mechanisms underlying each type. The work of Hécaen et al. exemplifies the modern neurological approach, in which calculation is analyzed into its component processes, specific types of acalculia are derived from the nature of the errors that are characteristic of such patients, and an attempt is made to systematically relate those different types of acalculia to particular cortical regions. The classification of acalculia into three types, presented below, continues to have a strong influence on the study of disorders of arithmetic, and many investigators still employ this scheme with very little modification:

Type 1. Acalculia resulting from alexia and agraphia for numbers, in which the patient is unable to read or write the numbers required for successful calculation. Although this form of acalculia has been referred to as *aphasic acalculia* (Benson & Weir, 1972), it is not limited to aphasic patients. This form of disrupted calculation may occur independent of an inability to read or write linguistic material and has been correlated mainly with posterior left (and sometimes bilateral) cerebral lesions (Hécaen, 1962).

Type 2. Acalculia of the spatial type is associated with impaired spatial organization of numbers, such as misalignment of digits in columns, inversions (6 for 9), reversals (12 for 21), visual neglect, and difficulties maintaining the decimal place. This type of acalculia is believed to be produced by posterior right-hemisphere damage or dysfunction. Hécaen et al. (1961) found this type of acalculia to be 12 times more frequent in right- versus left-hemisphere lesions.

Type 3. Anarithmetria refers to a disruption of calculation per se. This would be considered a primary acalculia within the Berger (1926) dichotomy; it refers to an inability to carry out arithmetic procedures despite intact visual–spatial skills and the capacity to read and write numbers. As with acalculia secondary to alexia and agraphia for numbers, anarithmetria was found to be predominantly associated with posterior left or bilateral lesions. However, approximately 20% of these patients had right-hemisphere lesions, a finding that emphasizes the difficulty of attempting strict localization of calculation ability without regard to the presenting phenotype.

The research of Hécaen and his colleagues has been of considerable heuristic value in the study of brain–behavior relationships in calculation. Their classification system and clinicopathological correlations led to many testable propositions that have been the basis of numerous detailed studies of the relationship between acalculia and other neurological and neuropsychological impairments (e.g., Dahmen, Hartje, Büssing, & Sturm, 1982; Grafman, Passafiume, Faglioni, & Boller, 1982). Furthermore, many of the basic concepts contained in the works of Henschen, Berger, and Hécaen have proved to be essential ingredients in the identification and classification of developmental disorders of calculation. For example, these authors pointed out that disordered arithmetic is not a univocal phenomenon but, rather, can result from the disruption of quite different underlying mechanisms.

The various sources of disordered arithmetic performance may involve calculation per se, one or more of its component processes, a prerequisite skill, or even a generalized impairment of which disordered calculation is merely a secondary symptom. All of these ideas have extended to the study of arithmetic LD, whether as part of a conceptual framework guiding the interpretation of empirical data or as premises in arguments from analogy. The latter point has particular import for the study of arithmetic and related LD, because much of the theorizing in this area has relied heavily on inferences made from studies of adults (Semrud-Clikeman & Hynd, 1990). Only rather recently has a substantial body of child/developmental data and data on model-building with respect to arithmetic LD begun to emerge (e.g., DeLuca, Rourke, & Del Dotto, 1991; Rourke, 1982, 1987, 1988, 1989;

Rourke & Fisk, 1988; Share, Moffitt, & Silva, 1988; White, Moffitt, & Silva, 1992).

Gerstmann Syndrome

Josef Gerstmann published a series of articles from 1924 to 1930 that described a constellation of four behavioral deficits that were reported to appear together as a syndrome. These deficits include the following: bilateral finger agnosia (inability to identify one's fingers by touch alone), right–left confusion, dysgraphia (disrupted ability to write), and dyscalculia. There does not appear to be any specific type of calculation disturbance that is characteristic of the Gerstmann syndrome (Hartje, 1987). Developmental Gerstmann syndrome (DGS) has also been proposed, and a fifth symptom, constructional dyspraxia, is often included in this classification (Kinsbourne, 1968; Kinsbourne & Warrington, 1963). According to Gerstmann (1940), the aggregate appearance of these deficits was related to focal damage or disease in the territory of the angular gyrus of the dominant (usually left) hemisphere.

Subsequent investigations, however, revealed that these deficits do not necessarily always appear together. Rather, each of them may appear in isolation, or they may appear as partial groupings in which only two or three are present in a particular individual (Heimburger, DeMyer, & Reitan, 1964). Furthermore, it was found that patients with all four Gerstmann symptoms invariably had large lesions involving the superior temporal and supramarginal gyri as well as the angular gyrus. Heimburger et al. found no case of full Gerstmann syndrome in which damage was limited to the angular gyrus, and in 3 of 23 such patients the angular gyrus was not involved at all. In addition, some patients with left angular gyrus lesions showed no Gerstmann symptoms. These clinico-anatomical comparisons are consistent with statistical analyses revealing that these deficits are no more strongly correlated with each other than they are with such deficits as poor visual memory, dyslexia, or constructional dyspraxia (Benton, 1961).

Consequently, it has been suggested that the Gerstmann syndrome is an artifact of biased observation and should not be considered a true clinical syndrome. However, damage or dysfunction in the parietal–occipital region of the language-dominant hemisphere does seem to be associated with the behavioral deficits described by Gerstmann, and the utility of the label persists despite its questionable status as a distinct diagnostic entity (Gaddes, 1985). More recent reports have presented cases that fit the syndrome and its proposed anatomical basis quite well. Although the syndrome remains largely an enigma, it appears to have some heuristic value and may, in fact, be a more common manifestation of developmental disabilities than was previously thought to be the case (Benton, 1992; Grigsby, Kemper, & Hagerman, 1987; PeBenito, Fisch, & Fisch, 1988; Spellacy & Peter, 1978). For example, Grigsby et al. reported on a group of children with Fragile X syndrome who exhibited three or more Gerstmann symptoms without any evidence of aphasia, as well as one boy with Fragile X who exhibited all five DGS symptoms. They concluded that DGS is in fact a clinical entity, and that a variety of partial symptom groupings from this syndrome are relatively common in children with Fragile X syndrome. It should also be noted that these symptoms form part of the syndrome of nonverbal learning disabilities (NLD; Rourke, 1989), the neuropsychological assets and deficits of which have been shown to characterize many forms of pediatric neurological disease, disorder, and dysfunction (Rourke, 1995).

Gerstmann syndrome is included in the present discussion not only because dyscalculia is one of its defining features, but also because it represents one of the first examples of a neuropsychological description of arithmetic LD. In fact, children with developmental Gerstmann's syndrome bear more than a passing resemblance to a subtype of children with arithemetic disabilities identified by Rourke and his colleagues (Rourke & Finlayson, 1978; Rourke & Strang, 1978; Strang & Rourke, 1983) using a developmental neuropsychological approach. As we will see, this approach raises some interesting questions regarding the differential specialization of the cerebral hemispheres in general, as well as the neuropsychological bases of arithmetic and mathematical reasoning in particular. In addition to providing a developmental model of central processing deficiencies in children, these studies have demonstrated that deficient performance in arithmetic can result from vastly different patterns of neuropsychological assets and deficits (Rourke, 1993).

Relevance of Adult Studies

There is, of course, some question as to whether knowledge obtained from the study of adults generalizes well to brain–behavior relationships in children. There is little doubt that analogies drawn between childhood and adult syndromes are conceptually useful and may provide a first step in the development of clinical classifications (Denckla, 1973). However, in many respects, the behavioral manifestations of brain damage in adults differ quite dramatically from those seen in children. This is a consequence of a number of interacting factors, including the nature of the damage, its location within the CNS, and the premorbid skills of the individual. For example, the type of damage most typically seen in adults differs from that seen in children. Focal intracerebral lesions resulting from cerebral vascular accident, tumor, or penetrating head injury are more common in adults, whereas more generalized impairments arising from perinatal trauma, anoxia, inborn errors of metabolism, or closed-head injury are typical of children. In short, brain disorders of childhood have been far less likely to be subject to strict anatomic delimitation, neuroimaging, neurosurgical intervention, or pathological inspection (Boll & Barth, 1981). However, this state of affairs is beginning to change dramatically for many brain diseases of childhood, including hydrocephalus (Fletcher, Brookshire, Bohan, Brandt, & Davidson, 1995) and traumatic brain injury (Ewing-Cobbs, Fletcher, & Levin, 1995).

Adults also bring a history of established function, learned skills, and accomplishment to the clinical or research situation. This history of premorbid development can result in a very different picture of brain–behavior relationships in adults than in children, who differ not only in terms of their skills and consequent strategies, but also in terms of the amount of change in these that is to be expected over time. Whereas adults exhibit relatively static brain–behavior connections, such relationships in children are of a much more dynamic nature. In children, the relevant issue is not only the loss or disrupted acquisition of specific skills, but also the impact that various neurodevelopmental impairments have on the order, rate, and level of future development and learning capacity. The effect that CNS damage or dysfunction will have on a child's arithmetic performance is very much a function of the child's current and future developmental demands as well as the neuropathological characteristics of the damage (Rourke, Bakker, Fisk, & Strang, 1983).

Consequently, consideration of the impact that various types of brain impairment may have on the developmental course of events may be more informative vis-à-vis brain–behavior relationships in arithmetic disabilities than via simply attempting to relate particular abilities to specific brain systems or regions. This is not to say that the lessons learned from a neurological approach to acalculia do not have relevance for the study of children with different subtypes of arithmetic LD: It is apparent that children so classified come by their difficulties with arithmetic for distinctly different reasons, and the relevance of different approaches to such disabilities depends on the specific nature of the subtype of arithmetic disability in question. It is clear that the conceptualization of some subtypes of arithmetic learning disability is greatly enhanced via comparison of those subtypes to documented cases of brain damage in both adults and children. However, it is also the case that some children exhibit a subtype of arithmetic LD that is better conceptualized in neurodevelopmental terms that bear little or no resemblance to neurological models of acalculia. It is precisely these distinctions that need to be addressed before detailed subtyping of arithmetic LD can proceed.

DEVELOPMENTAL DYSCALCULIA

Developmental dyscalculia has been relatively neglected compared to acalculia in adults or dyslexia in children. This is apparent in the absence of widely accepted criteria for its definition in research or its diagnosis in clinical settings. Kosc (1974) presented one of the most thorough discussions of this problem, with an emphasis on hereditary or congenital factors that may compromise the integrity of neural substrates of calculation ability. Based on evidence from a number of neurological, neuropsychological, and genetic studies, Kosc argued that developmental dyscalculia is properly considered to be a reflection of brain dysfunction, and defined it as follows:

> Developmental dyscalculia is a structural disorder of mathematical abilities which has its origin in a genetic or congenital disorder of those parts of the brain that are the direct anatomico-physiological substrate of the maturation of the mathematical abilities adequate to age, without a simultaneous disorder of general mental functions. (Kosc, 1974, p. 47)

This definition makes three essential points. First, developmental dyscalculia involves a specific impairment of mathematical abilities, within the context of normal general mental abilities. This is essentially the same point that is made by authors attempting to define the term *learning disability,* distinct from definitions of mental retardation or other general intellectual impairment. Second, developmental dyscalculia is defined and identified according to the relationship that exists between the child's current mathematical abilities and those that can be considered normal for his or her age. Only through a careful age-appropriate analysis of the child's assets and deficits can a significant and "pathological" impairment be discerned. Third, dyscalculia is a developmental affliction distinguished from acquired forms of acalculia occurring in adulthood. Thus, the term *developmental dyscalculia* is reserved for those disorders that have their origins in "hereditary or congenital impairment of the growth dynamics of the brain centers, which are the organic substrate of mathematical abilities" (Kosc, 1974, p. 48). This formulation suggests that the crucial impairment depends more on the developmental sequence of the acquisition and refinement of progressively more complex neurocognitive systems than on calculation per se. This notion stands in marked contrast to the notion of developmental dyscalculia as a static impairment of calculation centers in the brain.

In addition to identifying essential defining features, Kosc (1974) classified six subtypes of developmental dyscalculia. Four of these subtypes can be seen to bear a resemblance to the adult forms of acalculia described above; others appear to reflect uniquely developmental dimensions of the disorder. For example, Kosc described "verbal dyscalculia," in which there is a disruption of the ability to name mathematical terms and relations. These children have difficulty naming amounts and numbers of objects, operational symbols, and even digits and numerals. He also described a "lexical dyscalculia" (with impairment of the ability to read mathematical symbols, including digits, numbers, and operational signs) as well as "graphical dyscalculia" (in which the disability is manifested as a difficulty with writing numbers and operational symbols).

These patterns of impairment are similar to those reported by Hécaen et al. (1961) for adults with alexia and agraphia for numbers, in whom the functional integrity of the perisylvian regions, especially of the left hemisphere, is implicated. Kosc (1974) also described "operational dyscalculia," which is a direct impairment of the ability to carry out arithmetic operations per se. This form of developmental dyscalculia appears to be roughly equivalent to Hécaen's anarithmetria (Hécaen, 1962; Hécaen et al., 1961). However, it is unlikely that anatomical inferences from adult cases of anarithmetria are directly applicable to operational dyscalculia in children.

Kosc (1974) also proposed "practognostic" and "ideognostic" dyscalculias. Practognostic dyscalculia refers to a disturbance of the ability to manipulate real or pictured objects for mathematical purposes. This includes problems with enumerating a group of objects and estimating and comparing quantities. These children are unable to set out objects in order according to magnitude, show which of two items is bigger or smaller, or correctly indicate when two objects are the same size. Ideognostical dyscalculia is an impairment of the ability to understand mathematical ideas and relations required for mental calculation. Such persons may be able to read and write numbers but cannot understand what they have written. For example, the child might read the digit "9" but be unable to understand relations such as that 9 is half of 18, or is 1 less than 10, or is equivalent to 3 × 3 (Kosc, 1974).

It is interesting to note that the difficulties referred to as "practognostic" bear a striking resemblance to Piagetian tasks. A more recent study demonstrated that two 9-year-old boys who were unable to acquire elementary numerical skills and who exhibited deficits associated with Gerstmann syndrome had not progressed to the concrete operational stage of cognitive development (Saxe & Shaheen, 1981). Both children believed that changing the visual–spatial configuration of either a continuous or a discontinuous quantity actually changed its amount.

It would seem likely that practognostic and ideognostic forms of dyscalculia reflect fundamental impairments in, or failure to develop significant dimensions of, more basic concept-formation and nonverbal reasoning abilities. Neuropsychological studies of children with arithmetic LD, discussed below, have provided clues to a possible source of such atypical cognitive development.

CEREBRAL ASYMMETRY AND DISABILITIES IN ARITHMETIC

Understanding brain–behavior relationships in children who exhibit disabilities of arithmetic and mathematical reasoning requires at least a general familiarity with some issues surrounding cerebral asymmetry. It has been known for some time that the left and right cerebral hemispheres are not precise mirror images of each other; this applies to both their structure and their function. Each hemisphere has its own particular penchants, with some relatively straightforward lateralization of function being empirically demonstrable. The most well-known difference between the cerebral hemispheres is that the left hemisphere is usually dominant for language functions, whereas systems within the right hemisphere usually predominate in the processing of nonverbal stimuli. Such differences have been demonstrated in intact subjects using methods such as dichotic listening (Kimura, 1963), tachistoscopic stimulus presentations (Reuter-Lorenz, Kinsbourne, & Moscovitch, 1990), task-related EEG asymmetries (Doyle, Ornstein, & Galin, 1974; Earle, 1985; Rebert, Wexler, & Sproul, 1978), and average evoked potentials (Davis & Wada, 1977; Galin & Ellis, 1975; Licht, Bakker, Kok, & Bouma, 1992). Furthermore, these interhemispheric differences in function appear to have an anatomical basis. It has been reported that the left hemisphere tends to be slightly heavier and larger in most right-handed persons, with the largest differences being found in areas that mediate language functions (e.g., the planum temporale; Galaburda, LeMay, Kemper, & Geschwind, 1978; Geschwind & Levitsky, 1968).

Analyses of neuropsychological deficits arising from right- versus left-hemisphere lesions led some early investigators to speculate that the cellular organization of the left hemisphere is more close-knit and integrated than that of the right hemisphere (Hécaen & Angelergues, 1963). For example, visual–spatial deficits often result from lesions occurring over a broad range of areas within the right hemisphere, whereas deficits arising from left-hemisphere damage tend to be associated with more specific lesion sites (De Renzi, 1978; De Renzi & Faglioni, 1967). Furthermore, there is evidence to suggest that the overall functioning of the right hemisphere is more easily disrupted, even by relatively small lesions (Kertesz & Dobrowolski, 1981). Further support for this position comes from observations that tactile discrimination is more often disrupted bilaterally by right-hemisphere lesions than is the case with left-sided lesions, which tend to produce only contralateral tactile deficits (Semmes, 1968). As will be discussed later, this latter finding has some import for the neuropsychological profile analysis of one subtype of children with arithmetic LD.

More comprehensive theoretical accounts of these and other observations have been formulated by Goldberg and Costa (1981) and Rourke (1982). Goldberg and Costa incorporated anatomical and behavioral evidence into their theory of cerebral asymmetry, which holds that the left hemisphere is specialized for the processing of unimodal stimuli and routinized behavioral acts, whereas the right hemisphere is specialized for intermodal integration, processing of novel stimuli, and dealing with informational complexity. In particular, they pointed out that the structure of the left hemisphere is marked by the presence of three prominent opercula (clumps of gray matter) in the temporal, parietal, and posterior frontal regions, each of which appears to mediate relatively discrete and routinized functions (such as those involved in linguistic processes). Focal damage to one of these opercula tends to produce rather specific deficits, with other areas of the left hemisphere continuing to function in a surprisingly independent fashion. This arrangement can be contrasted with that of the right hemisphere, in which the prominent organizational feature is a higher ratio of white matter relative to gray matter, which appears ideally suited to the integration of complex information arriving through a number of sensory modalities. According to Goldberg and Costa, this results in an advantage; and a propensity for the processing of novel and/or complex stimuli, and this general organizational principle renders the right hemisphere more susceptible to generalized dysfunction arising from virtually any form of significant insult to its overall integrity.

Any attempt to relate arithmetic and mathematical ability to cerebral asymmetry must necessarily take into account the specific nature of the skill or ability under investigation. Phrenological notions of a single process or set of processes mediated by a calculation center in the brain have long since been abandoned and have yielded to more sophisticated accounts of how the brain might mediate these behaviors and abilities. Although the left hemisphere is generally believed to mediate the numerical symbol system, retrieval of number facts from semantic memory, and simple linear equations with an $a + b = c$ form (Geary, 1993; Lezak, 1983; Spiers, 1987), the right hemisphere undoubtedly plays an important role in mathematical performance that requires adaptive reasoning and/or visual–spatial organization of the elements of the problem (Rourke, 1993). Examples of the former would include the use of multiplication table values and story problems such as those found in the Arithmetic subtest of the Wechsler Adult Intelligence Scale–Revised (Wechsler, 1981). Examples of the latter type of task would depend partly on the strategy employed by the individual, but it is probably safe to assume that most persons draw on visual–spatial abilities during the procedures required for long division and multiplication.

In adults, mathematical performance appears to be mediated largely by the posterior association cortex, with left-sided lesions resulting in loss or disruption of basic arithmetic operations and number facts, including the concept of number itself. Right-sided lesions produce deficits in dealing with the visual–spatial–organizational dimensions of calculation and mathematical reasoning, such as using decimal places and "carrying" and "borrowing" (Grewel, 1952). However, this situation is greatly complicated by developmental dimensions and interactions. Superimposed on this already complex picture of brain–behavior connections for arithmetic performance is the added dimension of a developmental sequence of events that may become disrupted at a number of different points or stages. In turn, points in the developmental sequence of events probably differ in terms of the nature and location of CNS damage or dysfunction that will most negatively affect subsequent developmental events.

A NEURODEVELOPMENTAL APPROACH TO ARITHMETIC DISABILITIES

In a series of studies conducted from the 1970s to the present, Rourke and colleagues have described two subtypes of children with LD who exhibit equally impaired levels of arithmetic achievement but have vastly different profiles of neuropsychological assets and deficits (Rourke, 1993). Those studies were undertaken to determine the neuropsychological significance of different patterns of academic achievement. Previous studies examining Verbal IQ–Performance IQ discrepancies in children (Rourke, Dietrich, & Young, 1973; Rourke & Telegdy, 1971; Rourke, Young, & Flewelling, 1971) revealed predictable patterns of performance on Wide Range Achievement

Test (WRAT; Jastak & Jastak, 1965) Reading, Spelling, and Arithmetic subtests. The subsequent studies were conducted to determine if children with specific subtypes of LD, identified by their patterns of performance on the WRAT, would also demonstrate predictable patterns of neuropsychological assets and deficits (Rourke, 1993).

The first study in that series examined three groups of children with LD between the ages of 9 and 14 years who were equated for age and Wechsler Intelligence Scale for Children (WISC) Full Scale IQ (Rourke & Finlayson, 1978). Group 1 children were uniformly deficient in reading, spelling, and arithmetic; Group 2 participants performed significantly better (although still below age expectation) in arithmetic than in reading and spelling; and Group 3 children had markedly impaired arithmetic performance within a context of normal reading and spelling ability. All three groups exhibited impaired arithmetic performance, but only Groups 2 and 3 were equivalent, performing significantly better than Group 1 on the WRAT Arithmetic subtest. It is important to note that Groups 2 and 3 exhibited equally impaired levels of arithmetic performance, despite having very different overall profiles of achievement.

The results of the Rourke and Finlayson (1978) study indicated that Groups 1 and 2 performed significantly better than Group 3 on neuropsychological measures of visual–perceptual and visual–spatial abilities, whereas Group 3 performed significantly better on verbal and auditory–perceptual measures. Furthermore, children from Groups 1 and 2 exhibited a pattern of Verbal IQ < Performance IQ, whereas Group 3 children exhibited the opposite pattern, having lower Performance than Verbal IQs. These findings were interpreted as reflecting differential hemispheric impairment between the groups. That is, the findings were consistent with the hypothesis that children in Groups 1 and 2 had some impairment (i.e., relatively deficient functional integrity) of left-hemisphere systems, whereas Group 3 children exhibited the effects of compromised right-hemisphere functioning.

These inferences were based on the fact that subjects in Group 3 did particularly poorly only on those tasks thought to be subserved primarily by systems within the right cerebral hemisphere, whereas subjects in Groups 1 and 2 were deficient only on those tasks thought to be subserved primarily by systems within the left cerebral hemisphere. From this it was inferred that Groups 2 and 3, despite demonstrating equally impaired levels of arithmetic, differed in terms of the neuropsychological bases of those deficits. Group 2 children were apparently experiencing difficulties with arithmetic due to verbal deficiencies, whereas Group 3 children appeared to be encountering greater difficulty with the visual–spatial and nonverbal reasoning dimensions of arithmetic performance.

To explore the possibility that these groups were exhibiting differential impairment of right- versus left-hemisphere systems, Rourke and Strang (1978) examined the performances of these same three groups on measures of motor, psychomotor, and tactile–perceptual skills. The results indicated that Group 3 children were deficient, relative to both age norms and the performance of Groups 1 and 2, on complex psychomotor and tactile–perceptual skills, especially when using the left hand. This provided further evidence in support of the hypothesis that Group 3 children were experiencing their difficulties in arithmetic as a result of relatively deficient right-hemisphere systems, as opposed to Group 2 children, whose difficulties were apparently arising from compromised systems within the left hemisphere. Consistent with the present emphasis on arithmetic learning disabilities, the remainder of this review will focus on the performances of Groups 2 and 3 only. As previously mentioned, these two groups exhibited equally impaired levels of arithmetic achievement, apparently for very different reasons. The situation with respect to Group 1 is far more complex because it is probably the case that it is made up of a number of different LD subtypes (Fisk & Rourke, 1979).

A third investigation in the series (Strang & Rourke, 1983) compared the performances of children in Groups 2 and 3 on the Halstead Category Test (Reitan & Davison, 1974), a complex measure of nonverbal concept formation involving abstract reasoning, hypothesis testing, and the ability to benefit from positive and negative informational feedback. These adaptive dimensions of behavior, in addition to visual–spatial difficulties, were hypothesized to be instrumental in the arithmetic difficulties exhibited by Group 3 children. The two previous studies had demonstrated that Group 3 children exhibited a configuration of neuropsychological deficiencies that would have implications for their cognitive development in terms of Piagetian theory (Piaget, 1954). That is, these children's tactile–perceptual, psychomotor, and visual–perceptual–organizational deficiencies were seen as serious liabilities in terms of their being able to benefit from the early sensorimotor experiences that Piaget described as underlying the transition to later stages of cognitive development and acquisition of higher order cognitive skills. It is noteworthy that the participants in Saxe and Shaheen's (1981) study, who had not progressed to Piaget's concrete operational stage of cognitive development, exhibited neuropsychological profiles that were strikingly similar to those of Group 3 children.

As expected, Group 3 children made significantly more errors on the Category Test than did Group 2 children. Although the Halstead Category Test should not be considered a direct measure of right-hemisphere integrity, the development of higher order cognitive skills required for success on this measure is thought to be dependent on very basic developmental skills and abilities that appear to rely heavily on right-hemisphere systems for their successful elaboration (Rourke, 1989). That is, deficient performance on the Category Test was interpreted as reflecting a disordered pattern of development, and although this pattern was attributed to early neuropsychological deficits that appear to reflect relative dysfunction of systems within the right cerebral hemisphere, this does not imply that performance on the Category Test is lateralized to the right.

Based on the results of these three studies (see Table 3.1 for a summary), the following conclusions can be drawn regarding the neuropsychological significance of the two subtypes of children who exhibit arithmetic disabilities. First, at least two distinctly different patterns of neuropsychological assets and deficits can eventuate in arithmetic LD. Whereas Group 2 (now referred to as Group R-S) children exhibit normal levels of performance on visual–spatial–organizational, psychomotor, and tactile–perceptual tasks, Group 3 (now referred to as Group A, or the Nonverbal Learning Disabilities [NLD] subtype) children perform at impaired levels on these measures. Furthermore, children with NLD tend to encounter increasing levels of difficulty as the task demands become more novel and complex. In contrast, these children exhibit well-developed auditory–perceptual skills, especially for material that is amenable to rote verbal learning. Children of the R-S subtype have outstanding difficulties in these areas, especially with the complex semantic–acoustic aspects of the linguistic domain. It appears that Group R-S children encounter their difficulties with arithmetic as a result of verbal deficits that reflect relative impairment of left-hemisphere systems, whereas Group A (NLD) children are limited by nonverbal deficits that implicate relatively dysfunctional right-hemisphere systems.

Second, Group R-S children perform well on measures of nonverbal problem solving and concept formation. They exhibit intact capacities to benefit from nonverbal informational feedback, as well as from past experience with such tasks. This stands in marked contrast to the performance of NLD children, who exhibit outstanding deficits in these areas. This raises the question of whether NLD children are experiencing the cumulative effects of a disrupted sequence of developmental events. Their pattern of neuropsychological deficits may have affected early sensorimotor experiences, which in turn served to skew the normal course of cognitive development. The interested reader may wish to consult Rourke (1989, 1995) and Rourke and Fuerst (1995) for elaborations of the developmental dynamics of NLD children.

It is apparent that there is a need for further subtyping studies of arithmetic LD. As the above series of studies clearly demonstrates, a univocal

Table 3.1
Neuropsychological Assets and Deficits of Group R-S and Group A

Group	Assets	Deficits
R-S	Tactile perception	Auditory perception
	Visual perception	
	Psychomotor skills	
	Attraction to novelty	
	Tactile attention	Auditory attention
	Visual attention	Verbal attention
	Exploratory behavior	
	Tactile memory	Auditory memory
	Visual memory	Verbal memory
	Concept formation	
	Problem solving	
	Mathematical reasoning	
	Scientific reasoning	
	Linguistic prosody	Linguistic phonology
	Linguistic semantics	Verbal reception and repetition
	Linguistic content	Verbal storage
	Linguistic pragmatics	Verbal associations
	Linguistic function	Verbal output (volume)
A	Auditory perception	Tactile perception
	Simple motor skills	Visual perception
	Absorbing rote material	Psychomotor skills
		Aversion to novelty
	Auditory attention	Tactile attention
	Verbal attention	Visual attention
		Exploratory behavior
	Auditory memory	Tactile memory
	Verbal memory	Visual memory
		Concept formation
		Problem solcing
		Mathematical reasoning
		Scientific reasoning
	Linguistic phonology	Linguistic prosody
	Verbal reception and repetition	Linguistic semantics
	Verbal storage	Linguistic content
	Verbal associations	Linguistic pragmatics
	Verbal output (volume)	Linguistic function
		Verbatim memory

Note. For a full explanation of these neuropsychological profiles, see Rourke (1989, 1995).

conceptualization of arithmetic disabilities is simply not adequate to the task of understanding the unique assets and deficits of these children, or of developing adequate programs of intervention. Arithmetic LD can result from at least two very broad classes of neuropsychological impairment, one based on verbal deficiencies (probably reflecting relatively dysfunctional left-hemisphere systems) and one based on nonverbal deficiencies (which appear to reflect the phenotypical outcome of early impairment of, or lack of access to, systems within the right hemisphere). Evidence from studies of evoked potentials in children who exhibit these subtypes of LD strongly suggests that the conclusions with respect to relative hemispheric integrity are well founded. (For a review of those studies, see Dool, Stelmack, & Rourke, 1993.)

In view of the considerations alluded to above, each of these two subtypes of arithmetic LD may be divisible into more fine-grained subtypes. For example, in reference to the work of Hécaen et al. (1961) and Kosc (1974), it may be the case that children whose difficulties with arithmetic are attributable to verbal factors can be divided into two or more subtypes. That is, some of these children may have a variant of alexia and agraphia for numbers, whereas others may encounter difficulties with arithmetic secondary to more general linguistic deficiencies, difficulties retrieving number facts from semantic memory, or even discrete impairment of calculation per se. On the other hand, children whose difficulties with arithmetic are attributable to primarily nonverbal deficits may represent both a visual–spatial subtype and a nonverbal concept formation/adaptive reasoning subtype. However, in view of considerations regarding the marked tendency for right-hemispheral systems to be disrupted by significant impairment of virtually any locus, the latter would seem much more complicated than the formulations proposed by Hécaen and by Kosc. Other considerations with respect to a more fine-grained analysis of children whose disabilities in learning appear to be rooted in linguistic deficiencies are dealt with in Rourke (1989, chapter 8). The important points to emphasize at this juncture are that advances in the characterization of brain–behavior relationships in children with arithmetic LD are dependent on a more precise specification of the subtype of disordered arithmetic in question, and that this awaits further investigative effort.

SUMMARY AND CONCLUSIONS

As we have seen, the neurological approach to acalculia is, for the most part, concerned with localizing the particular component processes of arithmetic by correlating focal brain lesions with particular numerical deficits or types of errors. Mathematical performance is analyzed into its components, specific types of acalculia are derived, and an attempt is made to determine if these vary systematically with disease or dysfunction of particular cortical regions (Benton, 1987). The neurological approach has yielded a number of important inferences regarding the cerebral organization of mathematical abilities and, by analogy, these have informed efforts to elucidate relevant brain–behavior connections in children. It would appear that the component processes of arithmetic performance can be effectively dissociated (Geary, 1993; Spiers, 1987), and it is clear that clinico-anatomical correlations have provided compelling evidence for the differential involvement of particular cortical regions in these component processes (Hartje, 1987; Hécaen, 1962; Hécaen et al., 1961; Keller & Sutton, 1991).

This neurological approach has considerable conceptual utility for the examination of developmental disorders. The neuropsychological study of LD, however, is concerned more with phenotypic levels (i.e., manifest patterns of academic performance) and the relationships of these to more basic neuropsychological assets and deficits. A neuropsychological approach to LD is oriented toward the full range of brain–behavior relationships that may interact with or affect the arithmetic learning situation. In this approach, a systematic attempt is made to relate brain systems to the different ways in which arithmetic learning may be impeded. This can involve something as specific as retrieval of number facts from semantic memory, or as general as concept formation, nonverbal reasoning, and adaptive problem solving. Developmental neuropsychological deficits may involve calculation per se, difficulties with the visual–spatial demands of arithmetic performance, or a developmental lag or disruption that alters the child's normal course through the Piagetian stages of cognitive development.

The neurological approach to acalculia has yielded inferences that may seem somewhat contradictory to the findings from neuropsychological studies of children. The most prominent of these is that studies of adults implicate mainly the left cerebral hemisphere as being more important for calculation ability, whereas studies of children indicate that the right- and left-hemispheral systems are crucial to the development and elaboration of skills and abilities relating to the learning of arithmetic and mathematics. There are two broad reasons why findings from the study of acalculia in adults differ from many of the inferences drawn from neuropsychological studies of children.

First, a distinction can be drawn between a discrete impairment of the cortical "function" of calculation, or even one or more of its component processes, and the more general context of arithmetic learning and performance. Undoubtedly, the cognitive and neuropsychological demands of executing learned calculation skills differ considerably from those of initial arithmetic learning and performance in children. Whereas the former would rely heavily on previous rote learning and the retrieval of number facts from semantic memory, the latter seems much more dependent on the early maturation of concept formation and adaptive reasoning skills. Thus, a distinction can be made between the brain–behavior relationships that are relevant to the execution of learned calculation skills and the brain–behavior relationships that are relevant to the initial appreciation of prerequisite concepts and problem-solving skills required for successful arithmetic learning.

Second, according to the models proposed by Goldberg and Costa (1981) and Rourke (1982, 1989), the integrative, complex, and novel dimensions of early mathematical learning and concept formation would be expected to draw heavily upon the resources of right-hemispheral systems. Only after successful initial learning would number facts and basic arithmetic procedures become sufficiently routinized to be executed primarily by left-hemispheral systems. Thus, elements of arithmetic that were once very novel, conceptual, and even visual–spatial in nature for the child become automatic in the adult, even to the degree that many so-called calculations are merely specific instances of fact retrieval from semantic memory. The prediction regarding brain–behavior relationships that emerges from this view is that early damage or dysfunction in either hemisphere will disrupt arithmetic learning in the child, with very profound effects to be expected from early right-hemisphere insults, whereas left-hemisphere lesions will predominate in the clinico-pathological analysis of acalculia in adults. Our review of the literature suggests that this is the case.

Finally, issues of prediction and intervention are of paramount importance in any examination of children with LD. In the past, early identification of children at risk for disabilities in arithmetic was markedly neglected. It is now clear that neuropsychological assessment can reveal patterns of assets and deficits in children that are predictive of later academic performance, including arithmetic. More specific predictions of the types of difficulties likely to be encountered by these children will inform efforts to develop adequate programs of intervention (Rourke & Tsatsanis, 1965).

Although a detailed discussion of intervention in disabilities of arithmetic is beyond the scope of this chapter, it is clear that the views expressed herein have practical implications for the management of these children. The efficacy of an intervention program cannot be adequately assessed if the children in such a program have differing needs resulting from vastly different neuropsychological assets and deficits. Intervention programs tailored to the specific needs of arithmetic disability subtypes (e.g., Rourke, 1989, 1995; Rourke & Tsatsanis, 1995) are amenable to empirical investigation of their effectiveness, thus allowing for the modification and continued development of such efforts. In terms of overall progress in the field of learning disabilities, it is this heuristic dimension that represents the unique value of the study of brain–behavior relationships in children, including those with problems in arithmetic calculation.

4. Educational Aspects of Mathematics Disabilities

SUSAN PETERSON MILLER AND CECIL D. MERCER

Students with learning disabilities (LD) frequently have difficulty with mathematics computation and problem solving. Problems with mathematics usually begin in elementary school and continue through secondary school into adulthood. The statistics regarding math performance among students with learning disabilities are alarming. Cawley and Miller (1989) reported that 8- and 9-year-olds with learning disabilities performed at about a first-grade level on computation and application. Likewise, Fleischner, Garnett, and Shepherd (1982) found that sixth graders with learning disabilities solved basic addition facts no better than third graders without disabilities. These same researchers also found that fifth graders with learning disabilities solved one third as many multiplication problems as their peers without disabilities on timed assessments.

As Cawley, Baker-Kroczynski, and Urban (1992) reported, research has demonstrated that secondary students with mild disabilities attain math proficiency at the fifth- to sixth-grade level and that they perform poorly on required minimum competency tests. Similarly, Cawley and Miller (1989) reported that the mathematical knowledge of students with learning disabilities tends to progress approximately 1 year for every 2 years of school attendance. Warner, Alley, Schumaker, Deshler, and Clark (1980) found that adolescents with learning disabilities reached a mathematics plateau after seventh grade. The students in their study made an average of 1 year's growth during Grades 7 through 12. The mean math scores of 12th-grade students in both studies (Cawley & Miller, 1989; Warner et al., 1980) was high fifth grade. In another study, Greenstein and Strains (1977) found that adolescents with learning disabilities plateaued at the fourth-grade level and did not progress to higher stage problem solving.

Other researchers have spent time comparing the mathematics performance of students with disabilities with that of students without disabilities. Although reports have emerged discussing the poor math performance of many general education students in the United States (e.g., Dossey, Mullis, Lindquist, & Chambers, 1988; Lapointe, Mead, & Phillips, 1989), numerous investigators have found that students with learning disabilities experience even greater difficulty in math than their peers without disabilities (Ackerman, Anhalt, & Dykman, 1986; Fleischner et al., 1982; Goldman, 1989; Lee & Hudson, 1981; McLeod & Armstrong, 1982).

Many factors contribute to the poor math performance of students with disabilities. The purpose of this chapter is to provide a discussion of these factors. National reform movements are discussed first; then, learner characteristics are examined. Next, issues related to math instruction are presented. Finally, recommendations for improving math education for students with learning disabilities are shared.

NATIONAL REFORM MOVEMENTS

Several mathematics reform movements have emerged over the past 90+ years. From 1900 to 1935, the focus of math education was on solving problems of everyday life. Critics of this approach to mathematics learning believed the focus was too narrow and advocated meaningful arithmetic and problem solving using discovery methods (Carnine, 1992). During the 1950s, a sense of competition existed between the United States and the Soviet Union. Both countries were striving to advance their technological and academic abilities to establish themselves as a strong world power. In 1957 the Soviet Union beat the United States into space with the launching of Sputnik. This event resulted in increased concern among Americans regarding the math performance of school-age students. Consequently, federally funded experimental programs were introduced during the next two decades. Most of these programs, identified as "new math" or "modern math," continued to emphasize guided discovery of mathematical structures, patterns, and relationships. By the mid-1970s, another reform movement emerged that called for going "back to basics" (Sovchik, 1989) because students were unable to perform the basic operations. Unfortunately, none of these reform movements produced good results. Instead they neglected some of the basic psychological aspects of learning (e.g., attention, metacognition, memory, perception) and compounded the math problems of students with learning disabilities (Lerner, 1993). More recently, another reform movement (i.e., the National Council of Teachers of Mathematics [NCTM] Standards) has emerged that again advocates discovery learning via constructivism for teaching mathematics. Thus, the major reform efforts in math education appear to be cyclical (Carnine, 1992).

NCTM Standards

The National Council of Supervisors of Mathematics (1988) advocated for national reform in mathematics and issued their position on the essential components of a math curriculum. The NCTM (1989) endorsed and published these essential components in a document that has become known as the "NCTM Standards." The Standards document outlines what students should learn in mathematics during grades K through 4, Grades 5 through 8, and Grades 9 through 12. The Standards represent an attempt to change both math curricula and method pedagogy nationwide (Hofmeister, 1993). Generally speaking, the Standards represent increased expectations for student performance in mathematics. They emphasize that students must understand mathematical processes in order to communicate the language of math. Moreover, students should not memorize math without understanding the processes. The Standards also emphasize problem-solving skills and the importance of students' gaining "mathematical power" (NCTM, 1989, p. 5). Thomas Romberg, chairman of the commission that produced the Standards document, suggested that "mathematical power means having the experience and understanding to participate constructively in society" (Romberg, 1993, p. 37). The skills and concepts listed in the Standards are deemed appropriate for *all* students.

The Standards promote earlier instruction in advanced skills, such as algebra and geometry. The number of students taking algebra in the eighth grade is increasing. In 1990, estimates suggest, 16% of students took algebra in the eighth grade. That figure was up from the 13% noted in 1981 by the Second International Mathematics Study (Usiskin, 1993).

Numerous educators have expressed concern regarding the application of the Standards to students with disabilities (Carnine, 1992; Hofmeister, 1993; Hutchinson, 1993; Mercer, Harris, & Miller, 1993; Rivera, 1993). Among the concerns are the lack of references to students with disabilities in the Standards document, lack of research related to the Standards, and overall vagueness of the document. These issues need to be addressed if we are to avoid another failed reform movement, with students paying the greatest price.

Minimum Competency Testing

Another movement designed to increase expectations for student performance in mathematics has been the implementation of minimum competency testing. There has been a significant amount of controversy over the administration of competency tests. Disagreement has centered on the competencies that should be measured, how they should be measured, and whether they should be measured at all. In spite of this controversy, the trend of requiring testing for graduation has gained much momentum. Today, more than three fourths of the states require high school students to pass minimum competency tests before receiving a diploma (Lerner, 1993). The strongest supporters of minimum competency testing have often been businessmen, who view such testing as a method for promoting accountability in public education (Cohen, Safran, & Polloway, 1980).

Minimum competency tests are intended to ensure that before receiving high school diplomas, students can demonstrate at least minimal functional literacy skills, and, in some states and districts, students must demonstrate the ability to successfully apply basic skills to everyday life situations. The tests typically assess mathematical computation, reading comprehension, and writing. In some states, the tests also measure speaking, listening, reasoning, problem solving, reference skills, and life coping skills (Masters, Mori, & Mori, 1993). Students who are unable to pass the exam sometimes have access to an alternative diploma or certificate (e.g., certificate of attendance, special diploma).

Graduation Requirements

Another approach to increasing expectations of student performance in mathematics has been to increase the number of math courses students are required to take prior to graduation. In addition to requiring more units of math for high school graduation, some states are stipulating that higher level math courses, such as algebra, must be taken. To date, Louisiana, Mississippi, North Carolina, and a number of school districts, including Washington, DC, have established algebra as a graduation requirement. "Algebra for all" is becoming a frequently stated goal in school reform literature and is supported by the NCTM (Chambers, 1994). The trend toward higher standards in general education subject matter programming is continuing, as is evidenced by the increased reading difficulty of school textbooks, greater emphasis on problem-solving curricula, and the overall introduction of more rigorous requirements in secondary classroom assignments (Resnick & Resnick, 1985). These changes are making it more difficult for all students to meet the requirements for obtaining a standard high school diploma.

Inclusion Movement

Much has been written recently about the inclusion movement, or the concept that students with disabilities should receive their academic instruction in general education classrooms. The inclusion movement coincides with the trends, discussed previously, related to increased academic standards, competency testing, and more stringent graduation requirements. One of the primary goals of inclusion is to enhance students' social competence and to change the attitudes of teachers and students without disabilities (Gartner & Lipsky, 1987). Snell (1991) identified the development of social skills, the improvement of attitudes toward students with disabilities, and the development of positive relationships and friendships as the three most important benefits of integration. Given these potential benefits, it makes sense that many of the strongest supporters of full inclusion are advocates for students with severe disabilities (Fuchs & Fuchs, 1994). However, Roberts and Zubrick (1992) found that integrated students with disabilities were less frequently accepted and more frequently rejected than their peers without disabilities in general education classrooms.

With regard to academic progress (a primary concern for students with learning disabilities), other educators (Carnine & Kameenui, 1990; Fuchs & Fuchs, 1988; Hallahan, Keller, McKinney, Lloyd, & Bryan, 1988; Schumaker & Deshler, 1988) have reported that mainstreaming has not resulted in a high level of academic effectiveness. A 5-year longitudinal study involving more than 500 adolescents with learning disabilities revealed that these students were significantly more likely to fail in general classes than in special classes. The likelihood of failure increased relative to the number of general classes the students took and the length of time they spent in those classes. Most of the students in that study were graded against the same standards as their peers without disabilities and did not receive tutoring assistance. Moreover, the general education teachers in that study reported that they received little support in teaching the students with learning disabilities (Wagner, 1990). Other investigators (Semmel, Abernathy, Butera, & Lesar,

1991; Silver, 1991; Simmons, Fuchs, & Fuchs, 1991) concurred with Wagner, noting that few modifications are made for students with disabilities who are placed in general classrooms. Mather and Roberts (1994) suggested that the concept of a continuum of services for students with learning disabilities is threatened by the notion that these students should receive their instruction in general classrooms. They also mentioned the limitations (e.g., curricular materials, class size, knowledge of specialized teaching strategies) of general education settings in meeting the needs of students with disabilities. Therefore, it is not surprising that nonsupporters of full inclusion seem to be advocating for students with more mild disabilities who have higher academic goals (Fuchs & Fuchs, 1994). To accurately assess the effectiveness of teaching math to students with disabilities in general classrooms, adequate support systems need to be in place. Otherwise, students and teachers are predisposed to failure.

Learner Characteristics

Although each student is unique, an examination of the general characteristics of students with learning disabilities alerts teachers and researchers to learning factors that deserve attention when planning and teaching mathematics. Attributes of LD, information-processing factors, and characteristics in the areas of language, cognition, metacognition, and social and emotional behavior are worthy of inspection.

Attributes of LD

Many students with disabilities have histories of academic failure that contribute to the development of learned helplessness in math (Parmar & Cawley, 1991). It is postulated that learned helplessness in math results from youngsters repeatedly trying to solve problems when they have little or no understanding of mathematical concepts (e.g., students may practice computing division facts but do not understand what division means). This lack of understanding fosters the student's dependency on the teacher and thus promotes the belief that external help is needed to solve problems correctly. This cycle also helps create *passive learners,* a term that frequently is used to describe students with learning disabilities and refers to students who typically do not actively participate in or self-regulate their own learning (Parmar & Cawley, 1991). It is not surprising that many of these students are characterized as having motivational deficits due to their passivity.

Information-Processing Factors

The information-processing model provides numerous perspectives for examining the math difficulties of students with learning disabilities. Information-processing theory focuses on which information is acquired and how. Its primary features include attention, sensation, perception, short-term memory, long-term memory, and response (Bos & Vaughn, 1994). Students with learning disabilities frequently exhibit problems that contribute to poor math achievement and that are germane to information processing. Among these problems are attention deficits (Strang & Rourke, 1985; Zentall & Ferkis, 1993), memory problems (Strang & Rourke, 1985; Zentall & Ferkis, 1993), visual–spatial difficulties (Garnett, 1992), auditory-processing difficulties (Smith, 1994), motor disabilities (Smith, 1994), and information-processing deficits (Torgesen, 1990). The following are ways in which weaknesses in selected components of information processing may affect math performance:

Attention deficits:
1. Student has difficulty maintaining attention to steps in algorithms or problem solving.
2. Student has difficulty sustaining attention to critical instruction (e.g., teacher modeling).

Visual-spatial deficits:
1. Student loses place on the worksheet.
2. Student has difficulty differentiating between numbers (e.g., 6 and 9; 2 and 5; or 17 and 71), coins, the operation symbols, and clock hands.
3. Student has difficulty writing across the paper in a straight line.
4. Student has difficulty relating to directional aspects of math, for example, in problems involving up–down (e.g., addition), left–right (regrouping), and aligning of numbers.
5. Student has difficulty using a number line.

Auditory-processing difficulties:
1. Student has difficulty doing oral drills.
2. Student is unable to count on from within a sequence.

Memory problems:
1. Student is unable to retain math facts or new information.
2. Student forgets steps in an algorithm.
3. Student performs poorly on review lessons or mixed probes.
4. Student has difficulty telling time.
5. Student has difficulty solving multistep word problems.

Motor disabilities:

1. Student writes numbers illegibly, slowly, and inaccurately.
2. Student has difficulty writing numbers in small spaces (i.e., writes large).

Cognitive and Metacognitive Characteristics

In addition to the general learner characteristics discussed in the preceding section, students with learning disabilities also have difficulty with cognitive and metacognitive processes. Students who lack awareness of the skills, strategies, and resources that are needed to perform a task and who fail to use self-regulatory mechanisms to complete tasks will undoubtedly have problems with mathematics. Specifically, these students are described as having difficulty in (a) assessing their abilities to solve problems, (b) identifying and selecting appropriate strategies, (c) organizing information, (d) monitoring problem-solving processes, (e) evaluating problems for accuracy, and (f) generalizing strategies to appropriate situations (Brownell, Mellard, & Deshler, 1993; Cherkes-Julkowski, 1985; Goldman, 1989; Montague, chapter 9 of this book).

Several researchers (Kulak, 1993; Montague & Applegate, 1993; Swanson, 1990) have claimed that depicting students with learning disabilities as having metacognitive or cognitive deficits is only partially accurate. They note that many students attempt to use cognitive strategies, but the strategies they use may not be sufficient for solving the problem. For example, students may use numerous strategies with word problems (e.g., reading, checking, and computing strategies), but will not seem to have a working knowledge of strategies associated with representing problems (Hutchinson, 1993; Montague & Applegate, 1993; Montague, Bos, & Doucette, 1991; Zawaiza & Gerber, 1993). Problem representation involves converting linguistic and numerical information (via paraphrasing, visualizing, and hypothesizing) into mathematical equations and algorithms. Many students with disabilities find this task very difficult.

Language Disabilities

Because math symbols represent a way to express numerical language concepts, language skills become very important to math achievement. The use of language is requisite for calculations and word problems. In computing, language skills are needed to systematize the recall and use of many steps, rules, and math facts. For example, in the problem 73 x 96, there are about 33 steps involved (Strang & Rourke, 1985). The reading demands of word problems increase in each grade level. Irrelevant numerical and linguistic information in word problems is especially troublesome for many students with learning disabilities (Englert, Culatta, & Horn, 1987). Moreover, many students with learning disabilities have reading difficulties that interfere with their ability to solve word problems (Smith, 1994).

Social and Emotional Characteristics

The affective domain also is recognized as an important variable in the math performance of students with learning disabilities. For example, it is believed that repeated academic failure frequently results in low self-esteem and emotional passivity in mathematical learning (Cherkes-Julkowski, 1985). The emotional reaction of some individuals to math is so negative that they develop math anxiety. This condition is believed to stem from a fear of failure and low self-esteem and causes students to become so tense that their ability to solve, learn, or apply math is impaired (Slavin, 1991). Confused thinking, disorganization, avoidance behavior, and math phobia are common results (Conte, 1991; Zentall & Zentall, 1983).

Perspectives on Learner Characteristics

An examination of the literature suggests that many individuals with learning disabilities have numerous characteristics that predispose them to mathematical disabilities. Although some students with learning disabilities have a learning disability only in mathematics, others have a combination of academic disabilities (e.g., reading disability and math disability). Lovitt (1989) noted that other disabilities contribute to failure in math; for example, reading, language, and handwriting disabilities can have a strong negative influence on math performance. The heterogeneity of students with learning disabilities is apparent within math disabilities; that heterogeneity becomes even more of an issue when students without disabilities, students at risk, and students with learning disabilities and mild retardation participate continuously in the same math lessons. Recently, Parmar, Cawley, and Miller (1994) found that students with learning disabilities and mild retardation perform differently and require differentiated instruction. Moreover, Kavale, Fuchs, and Scruggs (1994) reported that students with learning disabilities and low achievers have differential learning characteristics. Thus, the complexity of math disabilities and learner characteristics presents a tremendous challenge for educators who aspire to improve their students' math performance. To compound this difficulty, educational factors other than learner characteristics also play a significant role in educational outcomes.

CURRICULA AND INSTRUCTION

Another educational factor that undoubtedly contributes to poor math performance among students with disabilities is poor curricula and instruction (Baroody & Hume, 1991; Carnine, 1991; Cawley, Fitzmaurice-Hayes, & Shaw, 1988; Cawley & Parmar, 1990; Kelly, Gersten, & Carnine, 1990; Mercer, 1992; Scheid, 1990). Teaching in today's schools is undoubtedly a difficult task: Poorly constructed texts and materials, coupled with the increasingly diverse student population, results in many classroom challenges.

Texts and Materials

At the elementary and middle school levels, basal mathematics programs are frequently used to guide instruction. Basal programs typically include a sequential set of student math books with accompanying student workbooks. Placement and achievement tests are frequently included to determine whether students have mastered the material in one book that then allows them to move to the next book. Teacher guides with suggestions for the teacher are provided in basal series. The typical basal curriculum uses a spiraling approach to instruction; in other words, numerous skills are rapidly introduced in a single graded book. The same skills are reintroduced in subsequent graded books at higher skill levels. Basal instruction using this spiraling-curriculum approach is supposed to add depth to the math topics taught, but in reality the result seems to be superficial coverage of many different skills. Skill mastery is unlikely, because new skills are introduced too quickly in an attempt to "get through the book." The primary concerns regarding basal programs are the lack of adequate practice and review, inadequate sequencing of problems, and an absence of strategy teaching and step-by-step procedures for teaching problem solving (Wilson & Sindelar, 1991). Research has demonstrated that the basal approach to teaching mathematics is particularly detrimental to students who have learning difficulties (Engelmann, Carnine, & Steely, 1991; Silbert & Carnine, 1990; Woodward, 1991).

At the high school level, textbooks are typically used to determine the instructional math program for students. Studies report that 75% to 90% of classroom instruction is based on textbooks, and, in most cases, those books define the scope and sequence of the material being taught (Tyson & Woodward, 1989). Cawley, Miller, and School (1987) found that high school math teachers put more emphasis on solving the problems in the textbook than on solving novel or life-based problems. Such an approach to teaching reduces the likelihood of generalization and limits adequate development of the cognitive and metacognitive strategies needed by many students with learning disabilities.

The lack of appropriate math materials for teachers to use compounds the problem of poor curricula and instruction. Historically, the reading and language problems of students with learning disabilities have received more research attention than their math disabilities (Bartel, 1982). Unfortunately, this negatively affects teacher training in mathematics and the creation of effective educational materials, which in turn negatively affects student learning. A related problem is the lack of field-testing of educational materials. Sprick (1987) reported that only 3% of educational materials are field-tested with students prior to being published. Thus, most commercial math materials are sold without first being used with students to determine their effectiveness. A marketing survey found that the most important characteristic in the sale of math textbooks was the attractiveness of the art (Carnine, 1992). To compound these problems, most textbooks are not written by teachers or individuals who have been trained as educators (Carnine, 1992).

Student Diversity

Classroom teachers are required to provide instruction for diverse student populations and are held accountable for covering the prescribed curriculum in a manner that ensures that most students in the class learn the content. Thus, teachers must decide whether to cover the full curriculum or spend sufficient instructional time on *part* of the curriculum, so that the slower students learn at least some of what is expected. Many teachers make the choice to "go on" (Bos & Vaughn, 1994). The decision to "go on" when teaching mathematics to students with learning disabilities can produce devastating results: Because math is hierarchical (i.e., new skills build on previously learned skills), students who are moved through the curriculum without understanding the foundation skills will continue to experience failure.

Because many students with disabilities receive at least part of their math instruction in general classrooms, it is important to examine accommodations made by general education teachers. An extensive survey of content area teachers at the elementary, middle, and high school levels (Bos & Vaughn, 1994) revealed that elementary teachers were more willing to undertake individualized student planning than secondary teachers. General education teachers at the middle and high school levels thought it was unrealistic to make special plans for individual students with disabilities. The teachers in the survey stated that the heterogeneous makeup of their classes and the large number of students they had to teach made it difficult to vary instructional procedures (Bos & Vaughn, 1994).

COMBINED EFFECT OF EDUCATIONAL FACTORS

An examination of the combined effect of the educational factors discussed in this chapter reveals possible explanations for the poor math performance of students with disabilities. These students have characteristics that make learning math difficult. For many students these difficulties are severe enough to warrant special education services, but, with the inclusion movement, it seems that more and more of them are receiving their academic instruction in general classrooms. Meanwhile, the curricula in general education classes are becoming more rigorous, with higher expectations. So, students who are already struggling with mathematics assignments are subjected to higher level math content and are somehow expected to succeed without differentiated instruction and adequate support. The NCTM Standards, math competency testing, and increasing math requirements for high school graduation all promote the idea that higher standards will result in improved skills. To date, little evidence exists to indicate that this desired outcome is actually materializing. To the contrary: Investigators report that the increased intensity of testing programs and new standards for promotion and graduation have increased student retentions and the dropout rate (Gamoran & Berends, 1987; Slavin & Madden, 1989). Moreover, case studies of young adults with learning disabilities who dropped out of school have revealed that a primary reason for leaving school is the feeling that "further academic efforts would be anxiety provoking and humiliating" (Lichtenstein, 1993, p. 345).

For many students, the acquisition of a certificate of attendance or special diploma is simply not a strong enough motivator to stay in school. In fact, these alternatives for students who cannot achieve the higher standards may actually be "life-biasing" and extend their disability label beyond the school years. Such bias may hinder the young adult's acceptance into mainstream society (Cohen et al., 1980). Therefore, it is imperative that special educators closely examine what is happening to students with math disabilities and begin to advocate for practices that are in the best interests of these students.

RECOMMENDATIONS FOR IMPROVING MATH EDUCATION

Develop a Refining Rather Than Reforming Posture

The continuous development of new math reform movements perpetuates a reactionary mode among teachers and ultimately reduces the likelihood that a solid knowledge base will drive the curriculum and instruction within public school mathematics classes and postsecondary education settings. Unfortunately, individuals with learning disabilities frequently pay the highest price for this phenomenon. Enough knowledge has been gathered to move math educators into a "refining" rather than a "reforming" mode. Educators who remain focused on what is known to be effective practice and who refuse to be distracted by unvalidated approaches will create the most successful elementary, secondary, and college programs for their students.

Mercer et al. (1993) offered three recommendations for improving current reform movements and facilitating a refining attitude among educators. First, dogma must be replaced by a recognition of the contributions of various paradigms. In other words, if a particular instructional approach results in student success, it should be valued regardless of its paradigm affiliation. Second, it is important to adopt a teacher–learner verification strategy. Reform in mathematics should be guided by replicable, validated programs that demonstrate effectiveness with targeted populations. The treatments should be clearly defined and able to be implemented with reasonable effort and the resources that typically are available in educational settings. Third, the diversity of learners should be considered when planning reforms. Sensitivity to differences among learners is needed for the benefit of both students and teachers.

Accommodate Learner Characteristics and Needs

Formal and informal assessment procedures are available to determine a student's current level of math performance (Montague, chapter 9 of this book). Once the student's strengths and weaknesses have been assessed, his or her mathematical needs should be explored, related to both short- and long-term goals. Mathematics instruction can then be planned that will bridge the gap that typically exists between students' current math performance and the skills needed to meet future goals. For example, if a student plans to attend a 4-year college after completing high school, the high school math curriculum should address the skills that will enable him or her to meet college entrance requirements, in addition to the skills that provide the math foundation required by the college the student plans to attend.

In addition to fostering math skills that complement students' educational goals, teachers should provide math instruction that will prepare individuals with disabilities to function independently after completing their education. For example, students need to acquire money skills; time skills; measurement skills; and an ability to add, subtract, multiply, and divide in order to function effectively in daily living.

When considering the diversity among all students—with and without disabilities—it seems unrealistic to assume that one curriculum or one set of standards will adequately meet the math needs of everyone. Similarly, the idea that there is one best method for teaching math to all students is not likely to have positive consequences for individuals with learning problems (Carnine, 1992). Moreover, to assume that students with differing intellectual and cognitive abilities can all learn the same type of math within the same time frame is also unrealistic. Forcing all students to follow one designated curriculum is a vivid example of fitting students to the curriculum rather than fitting the curriculum to students. Plainly stated, such an approach violates the basic principles of special education. Students with disabilities often need "things" that differ from what schools typically provide (Howell, Fox, & Morehead, 1993)—most individuals with learning disabilities are going to need accommodations or modifications in texts, materials, assignments, teaching methods, tests, and homework (Bateman, 1992). Regardless of *where* students with learning disabilities are taught math, individualization is going to be needed to adequately address the impact of the specific math disability that emerges from each individual's unique learning characteristics.

Implement Practices with Research Support

In addition to designing a curriculum that takes into account learner characteristics and learner goals, teachers of mathematics must implement the curriculum using effective instructional techniques with research support. Fortunately, researchers and educators have begun to identify best practices in mathematics instruction. Educators now have access to state-of-the-art information related to instructional practices that improve math performance among students with disabilities. Mastropieri, Scruggs, and Shiah (1991) conducted an extensive literature search and located 30 studies that validated instructional techniques for teaching mathematics to students with learning disabilities. Included among those techniques were (a) implementing demonstration, modeling, and feedback procedures; (b) providing reinforcement for fluency building; (c) using a concrete-to-abstract teaching sequence; (d) setting goals; (e) combining demonstration with permanent model; (f) using verbalization while solving problems; (g) teaching strategies for computation and problem solving; and (h) using peers, computers, and videodiscs as alternative delivery systems.

In another review, Mercer and Miller (1992) reported the major components of effective math instruction. Although their review showed much overlap with the findings of Mastropieri et al. (1991), a few additional components were identified, including monitoring student progress on a frequent basis, teaching math skills to mastery, and teaching generalization.

Most recently, Dixon (1994) reported research-based guidelines for selecting mathematics curricula. He suggested that adoption committees consider six guidelines when determining whether various math materials or curricula accommodate diverse learners. First, the concept of "big ideas" should be evident. "Big ideas" refers to the idea of teaching *important* math concepts that will facilitate the greatest amount of knowledge acquisition across the content being taught. Important concepts are taught to mastery, rather than briefly covering numerous math skills superficially. Included among the concepts that are generally accepted as "big ideas" are the four basic operations (addition, subtraction, multiplication, and division); place value; fractions; estimation; probability; volume and area; and word-problem solving. Other concepts may be added to this list as more research is done in this area. Dixon's second guideline was that explicit strategies should be included in math instruction. Explicitly taught strategies that can be applied to a large number of problems will be the most helpful to students with disabilities. The third guideline for selecting math curricula was to look for mediated scaffolding. Scaffolding is a means for students to obtain support while performing new skills to ensure successful learning. The support, or scaffolding, is gradually reduced as the student becomes more proficient at the skill. Scaffolding is needed after teacher demonstration and modeling, but prior to independent practice. The fourth guideline is to look for strategic integration. After skills have been taught in isolation, opportunities should exist for students to perform various types of problems that have been integrated into single lessons. Many concepts in math seem to be similar but are really very different (e.g., adding and multiplying fractions). Integrating practice (i.e., having students practice addition and multiplication of fractions concurrently) allows students to discriminate between various problem types. The fifth guideline was to consider primed background knowledge. Because mathematics depends heavily on previously learned skills, it is important to ensure that prerequisite information is obtained prior to the introduction of new skills. Ideally, prerequisite skills should be introduced and reviewed for several days (or even weeks) before the introduction of new, more complex skills. Dixon's last guideline for selecting mathematics curricula was to consider review practices. Review opportunities should be (a) sufficient for obtaining fluency, (b) distributed over time, (c) cumulative as more skills are learned, and (d) varied to promote generalization.

Because instructional practices with research support have been identified, it is important for educators to incorporate these practices into their curricula for students with learning disabilities. Teacher training programs and inservice activities need to include this information to reduce the lag time that inevitably occurs between completed research and program implementation. In addition to using instructional procedures that have research support, educators must continue with a mathematics research agenda. As new educational practices (e.g., inclusion, NCTM Standards) emerge, their implementation should be research driven. A proactive approach on the part of special educators will enhance outcomes for students and will waste less time and money.

In summary, there are many challenges embedded in the educational aspects of mathematics disabilities. In addition to diverse learner characteristics, a variety of environmental factors influence the progress that is made with regard to teaching mathematics to students with learning disabilities. Researchers and teachers must continue to work together to determine which curricula and instructional practices will bring about the best results in the shortest amount of time. Educators must resist the temptation to adopt the latest math movement, reform, or fad when data-based support is lacking (Carnine, 1992, 1994; Jones, Wilson, & Bhojwani, chapter 8 of this book; Mercer, Jordan, & Miller, 1994). Moreover, they must carefully consider how to keep students with disabilities in school and motivated to reach their life goals. Sensitivity coupled with knowledge will help facilitate the decision-making process when designing curricula, providing instruction, and developing school policy.

5. Educational Assessment of Mathematics Skills and Abilities

BRIAN R. BRYANT AND DIANE PEDROTTY RIVERA

Other chapters in this book discuss the nature of mathematics and how skills are taught to students who have mathematics learning disabilities. The purpose of this chapter is to provide the reader with an analysis of the current state of mathematics assessment when applied to individuals who have learning disabilities. More than ever, educators and psychologists have the tools necessary to provide valid assessments of individuals' mathematics skills and abilities (Bryant & Maddox, in press). There are more technically sound commercial mathematics tests on the market than ever before (Hammill, Brown, & Bryant, 1992), and the body of literature on effective assessment practices is ever increasing (Rivera & Bryant, 1992). Such statements could not have been made as recently as 20 years ago, when Buros (1975) prefaced his *Mathematics Tests and Reviews* by noting skeptically, "It is my considered belief that most standardized tests are poorly constructed, of questionable or unknown validity, pretentious in their claims, and likely to be misused more often than not" (p. xv).

Mathematics assessment plays a valuable role in identifying students who have mathematics learning disabilities (Bryant, Taylor, & Rivera, 1994; McKinney, Osbourne, & Schulte, 1993; Vogel, Hruby, & Adelman, 1993); such assessment is important in documenting the effects of mathematics instruction in a remedial or special program (Case, Harris, & Graham, 1992; Miller & Mercer, 1993; Montague, Applegate, & Marquard, 1993; Whinnery

& Fuchs, 1993), identifying the strategies that students employ during math activities (Parmar, 1992), conducting research about the characteristics of students with math LD (Hammill et al., 1992; Montague & Applegate, 1993), and examining the technical characteristics of mathematics tests (Fuchs et al., 1994; Jenkins & Jewell, 1992; Sandler, Hooper, Scarborough, Watson, & Levine, 1993; Slate, Jones, Graham, & Bower, 1994).

It should come as no surprise that mathematics assessment has been inseparably tied to the curriculum reform movements affecting mathematics instruction (Hofmeister, 1993; Hutchinson, 1993; Mercer, Harris, & Miller, 1993; Rivera, 1993). The reason this relationship was formed is quite simple: Curricular changes affect a measure's content validity, which is achieved principally by examining the mathematics curricula and selecting evaluation content that represent those curricula (Salvia & Ysseldyke, 1995). Rather than presenting in-depth discussions of assessment and mathematics curriculum reform, we recommend that readers take the time to examine Kliebard's (1995) seminal work, *The Struggle for the American Curriculum,* and Cooper's (1985) *Renegotiating Secondary School Mathematics.* These two sources document school reform extensively and provide an excellent perspective on America's dynamic curriculum development.

Throughout this book, reference has been made to the most recent effort in mathematics reform (see Rivera [chapter 1] and Miller and Mercer [chapter 4] in this book for a discussion about mathematics reform), the Standards of the National Council of Teachers of Mathematics (NCTM), published in 1989. Current assessment practices have also been influenced by the NCTM reforms (Bryant & Maddox, in press; Bryant et al., 1994; Rivera & Bryant, 1992). This chapter addresses the nature of current assessment practices by (a) providing a historical account of mathematics assessment to demonstrate the evolution of mathematics testing, and (b) describing specific assessment strategies that have been used effectively in recent years.

HISTORY OF MATHEMATICS TESTING

In light of the NCTM Standards' call for multiple assessment procedures that parallel trends in general education assessment practice, it is helpful to examine exactly what those trends have been over the years. Thorndike and Hagen (1955) examined the history of psychological and educational measurement and identified three periods: a pioneering phase (1900–1915), a "boon" period (1915–1930), and a period of critical evaluation (1930–1955). We borrow from their work and present information here that covers three related periods: the Pioneering Period (adjusted to 1845–1915), the Proliferation Period (1915–1940), and the Refinement Period (post-1940).

PIONEERING PERIOD

Greene, Jorgensen, and Gerberich (1953) provided an excellent review of early achievement testing. They chronicled the Boston school system's abandonment of its traditional oral examination practices in 1845 and its implementation of written evaluations to assess academic achievement, including arithmetic proficiency. The practice of oral examination had been firmly entrenched since the time of Socrates; but by the mid-19th century, the Boston school system's evaluation resources were overwhelmed by the sheer numbers of students who were attending public schools, thus necessitating a shift to written assessment. The new Boston school testing policy was observed closely by Horace Mann, then Secretary of the Massachusetts Board of Education, who advocated for this new evaluation system when he noted the following:

1. It is impartial.
2. It is just to the pupil.
3. It is more thorough than older forms of examination.
4. It prevents the "official interference" of the teacher.
5. It "determines, beyond appeal or gainsaying, whether the pupils have been faithfully and competently taught."
6. It takes away "all possibility of favoritism."
7. It makes the information obtained available to all.
8. It enables all to appraise the ease or difficulty of the questions. (Greene et al., 1953, p. 23)

Two decades later, Europe was credited with providing the first objective measure of mathematics when George Fischer, an English schoolmaster, wrote and administered his own scale in 1864. Thirty years later in the United States, J. M. Rice laid the foundation for the development of objective measures by devising the first spelling test with stringent administration and scoring guidelines. He administered the test to students in numerous school systems, analyzed its results, and used the data to identify trends in teaching. What is not reported in most historical accounts is that Rice also examined arithmetic performance (Thorndike, 1912). In addition to using his more famous spelling test to examine student performance, Rice also constructed an arithmetic test, administered the scale to schoolchildren in the last five grades of 18 schools in seven cities, and used the results to draw inferences about the status of arithmetic instruction. Perhaps his work in arithmetic is less frequently cited in the literature because his research results were less controversial for arithmetic instruction than for spelling instruction, but his efforts are of significance to this historical account nonetheless.

Most readers are very familiar with Binet's classic work with Parisian schoolchildren that led to his and Simon's (1905) famous intelligence test, the 1905 Scale (later the Stanford-Binet). But Burt (1922) noted that Binet also had written a test of achievement that included an arithmetic component. Binet's achievement test obviously had little effect on evaluation practices when compared to the contributions of his mental-ability measure, but the presence of such a scale is noteworthy.

In 1908, C. W. Stone published an arithmetic reasoning test (he used the test in his study of sixth graders in 26 different school systems); and, shortly thereafter (in 1910 and 1911), S. A. Courtis published the results of his arithmetic research using a test that measured fundamental operations; reasoning; speeded reasoning; and speed in computing simple addition, subtraction, multiplication, and division problems (Thorndike, 1912). Clearly, the pioneers had laid a foundation upon which to build, but one suspects that they had little idea of the testing boon that was to occur over the course of the succeeding decades.

PROLIFERATION PERIOD

By 1915, testing had become a popular endeavor among psychologists and educators, and over the course of the next 25 years the field of mathematics testing grew enormously. Between 1915 and 1925, more than 40 tests were published (Gilliland & Jordan, 1924), yet the emergence of the new scales was viewed with some critical concern:

> Once the movement took root, many became interested. There was an orgy of test making. No definite philosophy and method of validation had yet been developed. The origin of many of these tests is shrouded in mystery; no data are available. For this reason many tests, presumably carefully validated, were next to useless. (Tiegs, 1931, p. 75)

A significant event of the Proliferation Period occurred in 1923 with the introduction of the Stanford Achievement Test, written by three of the period's leading test developers: Truman L. Kelly, Giles M. Ruch, and Lewis

L. Terman. This standardized test gave educators the opportunity to efficiently test whole classes of students in a number of subjects, including mathematics. This standardized test was easy to score and withstood rigorous analyses in the research literature.

Two additional significant contributions to arithmetic testing were made by Ruch, Knight, Greene, and Studebaker (1925) and Buswell and John (1926). Ruch et al. created the Compass Diagnostic Tests in Arithmetic, which included a five-step assessment of story problems. Students read a word problem and were administered items that (a) assessed comprehension, (b) asked the students to identify what information was provided in the passage, (c) asked the students to provide information that was being called for by the problem, (d) required students to estimate the answer, and (e) asked them to select the correct solution.

Buswell and John (1926) examined the arithmetic performance of more than 500 children in 12 schools and identified 33 computation strategies that systematically lead to errors. Their scale was designed to help examiners identify faulty computation strategies, which they called "bad habits," to facilitate suggestions for remediation. As a result of the researchers' efforts, evaluations could be conducted to assess arithmetic processes in a systematic way.

Burt (1922) reported that psychologists looked for reasons for poor arithmetic test performance in much the same way that they did for reading (i.e., by examining long- and short-term memory and abstract reasoning abilities). Burt's reflections on this practice remain popular today:

> But on the whole, I have found such purely psychological tests less helpful than intensive tests of particular arithmetic operations. Arithmetic as practiced in the ordinary elementary school may be regarded as little more than a huge bundle of specific habits and memories. Hence, with the poor arithmetician the problem for the teacher is often simply this: to find which particular habit or memory is not operating as smoothly and as automatically as it should. (p. 304)

During the Proliferation Period, the use of achievement tests and intelligence tests as comparative measures was introduced. Thorndike's (1929) remarks about this practice are intriguing, particularly when one considers them in the context of the aptitude–achievement discrepancy analyses that are at the heart of learning disability diagnoses today:

> When tests of achievement are used in connection with measures of capacity, the treatment of each pupil may be made even more fair and fruitful. The pupil gifted with a high degree of native capacity is then expected to do more and better work than the average. He is thus protected against habits of idleness and conceit which might result were he constantly praised for merely exceeding

pupils of inferior endowment. Pupils of meager native talents are similarly protected against rebukes, scorn, and discouragement for inferiority in gross attainments. Each pupil's work becomes healthier and more fruitful when appraised in terms of his own capacities. Objective measurement, then, increases the effectiveness of education by setting up standards of achievement in terms of varying capacities. (p. 294)

Other events during this 25-year span that led to more sophisticated test development procedures included (a) increased numbers of texts on psychological and educational measurement; (b) the growth and popularity of the *Buros Mental Measurements Yearbook,* which allowed ready access to test critiques; (c) the application of available statistical procedures (e.g., correlation, factor analysis) to evaluate the technical characteristics of tests; and (d) the creation of achievement tests by psychologists schooled in the art of test construction. All of these events contributed to the critical evaluation of tests and the application of technical adequacy criteria to new measures.

REFINEMENT PERIOD

By the end of the Proliferation Period, educators had become accustomed to the ease of standardized test administration and scoring, and were largely convinced that the results of such tests were valid. However, the growing numbers of available tests, many of which had questionable technical merits, caused some concern regarding the extent to which standardized testing was being implemented. These concerns resulted in a period of scrutiny, wherein existing tests and newly created measures were examined for their ability to do what they claimed to do (Thorndike & Hagen, 1955).

As a result of the call for closer scrutiny of test use in America, the American Psychological Association (APA) published its *Technical Recommendations for Psychological Tests and Diagnostic Techniques* (1954), a document that was endorsed by the American Educational Research Association (AERA) and the National Council on Measurement in Education (NCME). The content of the document provided the testing community with a set of guidelines for test construction and use. Shortly thereafter, AERA and NCME teamed to generate the publication *Technical Recommendations for Achievement Tests* (1955), which served as an extension of the APA document. In 1966, the two publications were merged under the auspices of the APA, and the new document, *Standards for Educational and Psychological Tests and Manuals* (APA, 1966), provided standards for test developers and test users. This document has been revised several times in the intervening years but continues to influence the way educators and psychologists develop and use tests.

During the Refinement Period, two significant events dramatically altered the way mathematics was perceived in this country. The first, World War II, resulted in a mobilization of the workforce and a mindset for making advances in mathematics and science. As an example, Smith (cited in Kliebard, 1995) asserted that physics and mathematics instruction should be reoriented so as to place "greater stress upon aeromechanics, aeronautics, auto mechanics, navigation, gunnery, and other aspects of modern warfare" (p. 115). Prior to the Second World War, mathematics instruction had focused mainly on students' mastery of basic computation (McNeil, 1977).

Several projects were initiated to address the need for better mathematics instruction. One of the earliest such projects was the University of Illinois Committee on School Mathematics, which began in 1951 and lasted 10 years. This project resulted in the development of a consistent, unified mathematics discipline that would help students self-discover basic principles and develop manipulative skills needed to problem solve (Goodlad, Stoephasius, & Klein, 1966).

The second event of historical importance occurred in 1957 with the launching of Sputnik by the Soviet Union. Suddenly, the United States could no longer see itself as the leader in scientific endeavors, even though several programs had been established to enhance students' mathematics abilities. Throughout the 1950s, a variety of mathematics work groups had appeared, including the School Mathematics Study Group (late 1950s, into the 1960s), the Greater Cleveland Mathematics Program (beginning in 1959), the Madison Project of Syracuse University and Webster College (begun in 1957), the Experimental Teaching of Mathematics in the Elementary School (begun in 1959), and the University of Maryland Mathematics Project (initiated in 1957; Goodlad et al., 1966). The work of these groups and the Sputnik launching lent a sense of urgency to, and led to a, dramatic change in mathematics instruction. Ragan and Shepherd (1971) noted several effects of this curricular reform:

> To some this meant that, for some unknown reason, mathematical knowledge of the past was being replaced by some strange new kind of mathematics. To others it meant that changes were occurring not only in what but in how mathematics was being taught in our schools. The latter was the direction that the reform movement of the 1960s took in the school mathematics program. (p. 327)

It could be argued that the curriculum reform efforts of the 1950s, and the subsequent fast-paced instruction, caused an increasingly large group of

American students to fall behind their peers. Perhaps not coincidentally, *learning disabilities* was officially coined as a term in the early 1960s. Early learning disabilities assessments focused on *underlying processes* that were presumed to be associated with weaknesses in various areas (e.g., reading, mathematics). For example, Johnson and Mykelbust (1967) related arithmetic problems to such factors as auditory receptive language disorders and auditory memory problems. Assessment instruments, most notably the Illinois Test of Psycholinguistic Abilities (Kirk, McCarthy, & Kirk, 1968), were used to identify mathematics learning disabilities by focusing on the processes underlying academic failure. Education programs responsible for "remediating" student deficits in these underlying processes became prominent, but later research (e.g., Hammill & Larsen, 1974) debunked the programs. Thus, attention to testing and teaching specific mathematics skills replaced assessment and remediation of underlying processes. In the late 1960s and early 1970s, Connolly, Nactman, and Prichett (1971) examined major mathematics textbooks and other curricula to identify the skills that were being taught at the time. Items were generated to test the constructs in the areas of mathematical content, operations, and applications. The KeyMath Diagnostic Arithmetic Test became one of the first major efforts to create a comprehensive measure of numeration, fractions, geometry and symbols, addition, subtraction, multiplication, division, mental computation, numerical reasoning, word problems, missing elements, money, measurement, and time.

During the years since the introduction of the KeyMath, many more mathematics tests and evaluation procedures have been introduced to evaluate children's math strengths and weaknesses. In addition, the APA standards for test construction have been updated and were used by Hammill et al. (1989, 1992) to develop the objective criteria in *A Consumer's Guide to Tests in Print.* In this text, these researchers developed a set of criteria for evaluating the technical characteristics of norm-referenced tests. With the help of a team of raters (assessment authorities throughout the United States) the authors included ratings of more than100 of the most popular tests used in special education.

A review of Hammill et al.'s (1992) ratings for mathematics tests and general achievement tests that have mathematics components and that are used in learning disability assessments demonstrates that mathematics tests have indeed been refined to the point where almost all of them pass minimum criteria for technical adequacy (i.e., they achieved at least a B rating) regarding normative data, reliability, and validity. Those tests include the Basic Achievement Skills Individual Screener (Psychological Corp., 1983), Mathematics; the Diagnostic Achievement Battery, Second Edition (Newcomer, 1990), Math composite, Math Calculation and Math Reasoning subtests; the Diagnostic Achievement Test for Adolescents, Second Edition (Newcomer & Bryant, 1992), Math composite, Math

Calculation and Math Problem Solving subtests; the Kaufman Test of Educational Achievement–Comprehensive Form (Kaufman & Kaufman, 1985), Mathematics composite, Mathematics Application and Mathematics Computation subtests; the KeyMath–Revised (Connolly, 1988), for which 14 of 17 subtests and composites received B ratings, and 3 received F ratings; the Peabody Individual Achievement Test–Revised (Markwardt, 1989), Mathematics; the Quick-Score Achievement Test, Arithmetic (Hammill, Ammer, Cronin, Mandlebaum, & Quinby, 1987); the Scholastic Abilities Test for Adults (Bryant, Patton, & Dunn, 1991), Quantitative and Mathematics composites, Math Calculation and Math Application subtests; and the Test of Early Mathematics Abilities, Second Edition (Ginsburg & Baroody, 1990), Math Quotient.

Finally, the Refinement Period has seen an evolution in how tests are used for educational purposes. While continuing to see a need for well-built norm-referenced instruments, writers on assessment are cautioning against their use in developing instructional plans (Hammill & Bryant, 1991; Parmar, Frazita, & Cawley, 1996; Salvia & Ysseldyke, 1995; Taylor, 1993; Taylor, Tindal, Fuchs, & Bryant, 1993). Because norm-referenced tests are built to identify an examinee's achievement levels as compared with those of a national sample, and because they sample achievement areas rather than comprehensively assessing them (i.e., a relatively small number of items rather than a comprehensive selection, are chosen to represent a particular construct), these tests are ill-used to plan instruction. Instead, they provide examiners with general indexes of strength and weakness. Assessment authorities have called upon educators to use a variety of assessment techniques to plan and monitor instruction. Unfortunately, recent research (Lopez-Reyna, Bay, & Patrikakou, 1996) has demonstrated that the call has not been entirely heeded, but it is not for the lake of well-thought-out assessment schemes. The remainder of this article describes several methods that have been adapted or created during the Refinement Period and that have been successfully used to evaluate mathematics functioning. Such assessment methods are in keeping with the NCTM Standards' call for a variety of assessment techniques to be used when evaluating students' mathematics proficiency.

ASSESSMENT STRATEGIES

During the Refinement Period, psychologists and educators adapted several assessment procedures that had been developed in earlier periods, and also created new strategies designed to identify levels of math performance and to understand why students compute as they do. Some of these procedures are specific types of evaluations (e.g., criterion-referenced tests, curriculum-based measurements), and some are approaches to assessment (e.g., portfolios, observations). Many of the more popular procedures are described here.

Portfolios

Rivera and Smith (1996) noted that considerable interest has been generated in portfolio assessment as a means of monitoring student learning and evaluating the effectiveness of instructional programs and decision making (e.g., Paulson, Paulson, & Meyer, 1991; Swicegood, 1994; Valencia, 1990; Wolf, 1991). Paulson et al. defined portfolio assessment as

> a purposeful collection of student work that exhibits the student's efforts, progress, and achievements in one or more areas. The collection should include student participation in selecting content, the criteria for selection, the criteria for judging merit, and evidence of student self-reflection. (p. 60)

Portfolio assessment in mathematics is purported to be a useful tool for monitoring student progress as it relates to curricular objectives and instructional methods, for focusing more on the way students respond to math problems than on their answers, for measuring student academic achievement and classroom learning more directly, and for helping teachers in their instructional evaluations (Swicegood, 1994; Wesson & King, 1992).

Rivera and Smith (1996) also noted that there is some disagreement in the literature on what exactly constitutes portfolio assessment (e.g., some researchers draw an analogy between educational portfolios and artists' portfolios; others note the differences between instructional and assessment portfolios; still others provide different descriptions of portfolio content and format). However, certain characteristics emerge frequently in portfolio descriptions; these are next discussed in terms of content selection and analysis.

Content Selection. The two primary purposes of portfolio assessment are to document student progress and to guide instructional decision making on a regular basis in a particular content area. Therefore, for assessment purposes, the portfolio content should (a) represent a valid reflection of the curricular goals, (b) be collected within a specified time frame, (c) include student work that is derived from different instructional techniques, and (d) represent a variety of situations within which work is generated (Rivera,

1994). For a student with math learning disabilities, the items selected for inclusion in the portfolio should relate to his or her Individualized Education Program (IEP; Swicegood, 1994). Portfolio assessment should include data that help students progress in the designated curriculum when specific instructional procedures have been implemented. Examples of data that may be contained in the portfolio include "raw data" (e.g., a set of story problems, math problems); "summarizing data" (e.g., inventories, norm-referenced tests, rating scales) for evaluation purposes (Valencia, 1990); and curriculum-based data (e.g., anecdotal notes, academic graphs; Wesson & King, 1992).

Analysis. Students with mathematics learning disabilities often lack the specific academic skills and cognitive strategies that are needed to learn efficiently. For these students, portfolio assessments that include examples of completed work samples can be analyzed to identify the specific strategies that were used to derive the answers. For example, a student's completed word problems could be examined with the following questions in mind:

- Is the answer correct or incorrect?
- What computational skills were demonstrated or lacking?
- What reading errors may have contributed to the incorrect solution?
- What syntactical errors were made?
- What strategy was used to solve the problem?
- What visual aids (e.g., tallies, pictures, graphs) were used?

Teachers could use various techniques (e.g., "think aloud," observation, and error analysis) to solicit answers to these questions; the resulting information could be documented in the portfolio and used frequently to monitor student progress. Samples yielded by various other approaches outlined in the remaining sections of this article could also be included in a student's portfolio as examples of a variety of mathematics behaviors.

Cautions and Considerations. Although portfolios can be of use in the identification of students' math strengths and weaknesses, we urge teachers to use portfolio assessment judiciously. Used inefficiently, portfolios can become little more than a storehouse of student papers. Rivera and Smith (1996) suggested that teachers need to develop a systematic approach to data analysis and linkage to the IEP. Additionally, teachers of students with learning disabilities would be well advised to heed Coutinho and Malouf's (1993) assertion that "relatively little is known about the use of this approach with students with disabilities, the implications of large-scale performance assessment programs for these students, and the role of special educators in implementing the approach" (p. 63). It is our belief that teachers with a strong assessment background can find value in portfolios; techers inexperienced in gathering and interpreting assessment data may find portfolios to be of limited value.

Criterion-Referenced Tests

Criterion-referenced tests are used to demonstrate student knowledge of specific content, unrelated to peer performance (Hammill & Bryant, 1991). Well-constructed criterion-referenced tests contain a sufficient number of items of a specific type to determine whether the student has grasped the math content (Bryant & Maddox, in press). An important first step in generating a criterion-referenced test is to examine the scope and sequence chart that accompanies the curriculum being taught. For each skill, items representing the content are created. If mastery is set at *equal to or greater than 75%*, then at least four items per skill are written; if *equal to or greater than 80%* is selected as the mastery level, at least five items are logically generated; and if *equal to or greater than 90%* is selected, 10 items are usually created. To help with a test's content validity, actual items from the school's textbook should be selected.

Mercer and Mercer (1993) noted that many students who exhibit math problems produce correct answers, but they do so using laborious procedures. Because mathematics fluency is important, rate of performance should be examined (Baroody & Ginsburg, 1991; Hasselbring, Goin, & Bransford, 1987; Resnick & Ford, 1981). Bryant and Maddox (in press) provided a procedure for generating an index of rate and accuracy. To generate such a score, the test is administered to several students who are known to be proficient in the content being measured, average time for completion is multiplied by 1.5, and the resulting figure is set as the fluency criteria. Thus, if the average completion time for the 10 items is 50 seconds, the mastery level might be set at "9 or 10 of the items correct within 1 minute 15 seconds" (i.e., 50 × 1.5 = 75 seconds). When one applies the 1.5 criterion, however, a particular student's concomitant conditions (e.g., dexterity problems) have to be considered. For example, the student who has a mathematics learning disability and a dexterity problem would be ill-served if a strict fluency standard were applied, given the associated motor problems that affect transcribing the answer onto paper. Unfortunately, students with motor involvement too often are subjected to the same rate demands as peers without such conditions.

Curriculum-Based Measurement

Curriculum-based measurement (CBM) is a validated version of curriculum-based assessment. CBM involves the ongoing measurement of student performance when compared to the curricular outcomes of the school system (Fuchs & Fuchs, 1988; Shinn & Hubbard, 1993; Taylor, 1993; Tucker, 1985). Using validated and reliable procedures, instructional progress through the curriculum is monitored to identify the effectiveness of mathematics intervention (Fuchs, Fuchs, Hamlett, & Allinder, 1989). Developers of curriculum-based measurement (a) select long-term goals, (b) measure behaviors, (c) implement standardized measurement methods, (d) employ decision-making rules that guide instructional evaluation, and (e) accommodate a variety of instructional methods (Taylor, 1993).

Fuchs, Hamlett, and Fuchs (1990) devised a computerized application of CBM procedures in spelling, reading, and mathematics. These applications assist teachers on a daily basis by (a) evaluating students' learning rate, (b) determining when instructional changes are needed, (c) monitoring the adequacy of instructional goals, and (d) comparing the effectiveness of interventions for individual students (Whinnery & Stecker, 1992). Such a data-based decision-making procedure provides the teacher with a tool to implement and monitor effective instructional programs (Deno, 1985; Fuchs, Fuchs, Hamlett, & Allinder, 1991). Fuchs et al. (1990) also noted that such an intervention approach can lead to increased expectations of student performance.

Figure 5.1 provides a visual depiction of how CBM works. A student was administered a test from *Basic Math* (Fuchs et al., 1990), and the results were plotted on a graph. The midpoints of the student's test performance over time were used to plot a progress line (solid line), which exceeds the anticipated progress line (dotted line). Thus, the data show that expectations for this student could be raised because performance exceeded expectations. The solid line could have been lower than the dotted line, meaning that performance did not meet expectations. In either case, new decisions would be made regarding instructions—decisions that likely would not be made without a data-based intervention approach.

Rivera, Taylor, and Bryant (1994–1995) reviewed the literature between 1990 and 1995 to identify how the NCTM Standards are influencing current mathematics assessment procedures. Of the 10 studies found in the literature, 9 used CBM procedures, demonstrating partial conformity to NCTM recommendations. One would be hard-pressed to find a more effective technique than CBM for directing, monitoring, and redirecting remedial efforts.

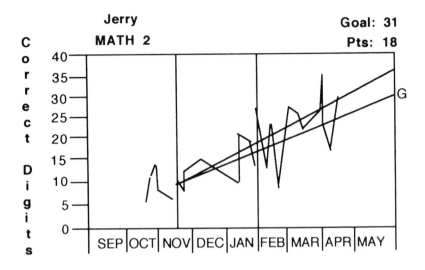

Figure 5.1. Sample student performance in *Basic Math*.

Calculation Error Analysis Using Structured Interviews and Checklists/Rating Scales

Baroody and Ginsburg (1991) noted that assessments would be incomplete without an error analysis that would identify the specific mistakes made by the examinee. Because mathematical errors can be interpreted as *incomplete responses* rather than *wrong responses* (DeRuiter & Wansart, 1982), it would be appropriate to borrow the word *miscue* from reading in identifying the incorrect strategies that are employed during problem solving.

An efficient way to identify the strategies students employ as they calculate is to conduct an interview (Bryant & Maddox, in press; Rivera & Bryant, 1992). Checklists and/or rating scales are useful tools for noting strategies that are discussed during the interview or observed during calculation. Generally, checklists take the form of dichotomous scales, wherein "yes/no" responses are provided; or Likert-type scales, with which the rater evaluates performance on a sliding scale that ranges from *never* to *always*. Table 5.1 provides an example of two different ratings using the same checklist, which is adapted from the work of Ginsburg and Mathews (1984).

In Table 5.1, student computational errors fall within two categories, *slips* and *bugs* (see Ginsburg, 1987). Slips are "minor deficiencies in the execution of a readily available procedure" (p. 435). For example, consider the student who knows how to regroup in addition and is asked to solve the problem 26 + 37 = ? The student correctly notes that 6 plus 7 is 13 but carelessly forgets to carry the 1 after writing down the 3. The remainder of the problem is then incorrectly solved by adding 2 plus 3, resulting in the answer of 53. In this case, the student knows better but has failed to apply the correct technique.

Bugs are systematic procedural errors that a student consistently makes during problem solving. Consider the student who overgeneralizes the "always subtract the smaller number from the larger number" rule. Here, the student's error is not random. Instead, he or she has applied a strategy because it is deemed correct. Such a bug usually results from the logical generalization of a known rule to a new situation. By using process assessment and classifying student errors according to number facts, slips, and bugs, valuable information is obtained that can lead to the development of an instructional program to remediate students' math deficiencies.

As noted, interviewing is a useful tool for conducting process assessments. Assessors usually select from three procedures to conduct an interview. In the first, the student is given a problem and asked to "think out loud" as the problem is being solved. During the interview, it is important to

Table 5.1
Sample Dichotomous and Likert-Type Checklist/Rating Scales: Subtraction Performance

	Dichotomous		Likert-type				
	Yes	No	Never				Always
Uses standard school method	1	0	0	1	2	3	4
Uses informal method	1	0	0	1	2	3	4
Aligns numerals correctly	1	0	0	1	2	3	4
Bugs							
Adds like multiplies	1	0	0	1	2	3	4
Zero makes zero	1	0	0	1	2	3	4
Adds from left to right	1	0	0	1	2	3	4
No renaming done: all digits on bottom	1	0	0	1	2	3	4
Renames wrong digit	1	0	0	1	2	3	4
Uses wrong operation	1	0	0	1	2	3	4
Adds individual digits	1	0	0	1	2	3	4
Other	1	0	0	1	2	3	4
Slips							
Skips numbers	1	0	0	1	2	3	4
Adds twice	1	0	0	1	2	3	4
Other	1	0	0	1	2	3	4

be mindful of Kennedy and Tipps's (1994) advice, "Do not interrupt during the explanation. Errors and possible reasons for misconceptions should be noted and used later to reteach the student" (p. 117).

Error Analysis in Word Problems

Bryant and Maddox (in press) suggested that as evaluators attempt to identify the strategies students use in solving story problems, they conduct a reading miscue analysis. Such an analysis can serve to identify reading behaviors that can interfere with problem solving (e.g., substituting incorrect words, omitting key words). For example, consider the word problem in Figure 5.2.

The student reads the passage silently first, then aloud; breakdowns that might be occurring as a result of decoding or comprehension are identified. In that example, several substitution miscues (words substituted for the text word; they appear above the text word) are made during oral reading.

The miscues represented in Figure 5.2 can easily affect the semantic and syntactic integrity of the passage (i.e., it is difficult to comprehend the passage when many vocabulary and grammar errors are made). Thus, it is possible that the problem will be answered incorrectly for reasons that have

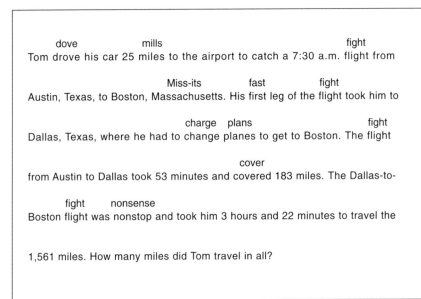

Figure 5.2. Sample miscue analysis for mathematics story problem evaluation.

nothing to do with computation or the ability to use computation in applied situations. To further check for understanding, several steps could be presented that are similar to those in the work done many years ago by Ruch et al. (1925). It is recommended that the student be allowed to refer back to the passage to find each answer, and perhaps use a highlighter to mark where each answer is found in the passage.

Step 1: Comprehension (Which statements are true?)
- Tom rode on an airplane all day.
- Tom traveled only to the airport.
- Tom flew on more than one plane.
- Tom traveled in more than one type of vehicle.
- Tom rode much farther in the car than in the plane.

Step 2: What Is Given (Select what is given in the story)
- Distance from home to the airport.
- How long it took to get to the airport.
- Where Tom flew to.
- The distance from Austin to San Antonio.
- The time Tom arrived in Boston.

Step 3: What Is Called For (Check what the problem is asking for)
- Distance traveled in the airplane alone.
- Distance from home to Dallas.
- Speed of the airplane.
- The total distance Tom traveled.
- The time it took to travel to Boston.
- Speed of the car.

Step 4: Estimation (What is probably the answer?)
- 80 minutes.
- About 500 miles.
- About 1,750 miles.
- 5 hours.
- About 3,000 miles.

Step 5: Solution (What shows the correct solution to the answer?)
- 25 + 730 + 53 + 183 + 322 + 1,561 = 2,874 miles
- 53 minutes + 3 hours 22 minutes = 4 hours, 15 minutes
- 25 + 183 + 1,561 = 1,769 miles
- 7 hours 30 minutes + 53 minutes + 3 hours 22 minutes = 11 hours 45 minutes

Table 5.2
Inventory of Mathematical Disposition Experiences

Student experiences	Date and activity	
Confidence in using mathematics for solving problems	1/15—Correctly solved 90% of problems assigned	2/12—Worked effectively as part of a small group that solved a problem
Flexibility in solving mathematics problems	2/2—Demonstrated different ways of solving a multiplication problem	3/18—Students challenged one another on solution methods
Perseverance in solving mathematics problems	1/29—Worked all day to collect and display data on favorite pizza topping	2/5—Worked diligently to identify ways to make change for $1—all ways found
Curiosity in performing mathematics activities	1/10—Solved a "what if" question and expressed answer in own words	2/7—Worked in small groups to generate own units for measuring room
Reflection on own thoughts	Each day students describe their thoughts during problem solving	
Value in applying mathematics to life activities	1/23—Each student brought in items for math application bulletin board	2.24—Field trip to science fair to see how mathematics applied
Appreciation of role of mathematics in life	1/25—Brought in magazine articles that contained mathematics term	2/14—Demonstrated place value system appreciation by finding sums using Roman numerals

Note. Adapted from CNTM (1989).

- 25 + 730+ 53 + 183 + 322 + 1,561 = 2,874 minutes
- 183 + 1,561 = 1,744 miles

After these questions are answered, the student could be asked to teach the steps that are taken during problem solving. Examiners could then check for faulty strategies and applications.

Observations

According to NCTM (1989), "Within the instructional context, teachers continually make informal judgments about their students' progress. Nonetheless, they often are reluctant to use these observations as the basis of important instructional decisions because of their potential subjectivity and unreliability" (p. 196). Observations provide valuable data that should be combined with the data accumulated via other assessment strategies to provide an overall assessment of the effectiveness of the instructional efforts.

There are numerous ways to observe students and make performance judgments. The NCTM Standards call for including subjective information on the disposition of students as they involve mathematics. Table 5.2 demonstrates how observations can be recorded for seven math disposition experiences. Here, an anecdotal record is used to provide information for seven targeted areas. Such a record allows for qualitative data gathering.

Student Perceptions

Observations also can be made from the student's perspective. It is sometimes helpful to gather information about a student's motivation and confidence level during an instructional activity (Baroody & Ginsburg, 1991; Cronin, 1985; DeRuiter & Wansart, 1982). Most students know whether they are successful or unsuccessful in a particular activity. Although students may be reluctant to share their insecurities initially, over time and in a supportive environment, that reluctance may diminish. One method to gauge students' perceptions is to ask them to complete a brief survey, such as the one presented in Figure 5.3, after they complete a math homework or seatwork activity. Such a scale provides an idea of how the student approached the exercise.

SUMMARY

In this chapter we have addressed the history of mathematics assessment and offered suggestions for assessment strategies that reflect students'

While doing this assignment, I felt (check one):

_____ confident that I knew how to solve all of the problems. I feel that I can teach others how to solve similar problems.

_____ like I knew how to solve some problems, but there were many that I did not feel sure about.

Explain

_____

_____

_____ like I thought I could solve the problems when I started, but then I got confused and couldn't remember how to solve them.

Explain

_____

_____

_____ lost from the start. I never understood what the teacher was doing during instruction.

Explain

_____

_____

Figure 5.3. Student self-report.

progress in the curriculum and possible defective cognitive strategies they employ during computation and problem solving. Mathematics assessment strategies have evolved over the years, keeping pace with curricular and instructional changes resulting from various reform efforts. In the field of learning disabilities, norm-referenced mathematics instruments play a vital role in helping professionals determine the presence of a learning disability and in developing a profile of the student's strengths and weaknesses. However, these instruments are not intended as tools for instructional planning. Rather, other math assessments practices, such as criterion-referenced testing, curriculum-based measurement, error analysis, clinical interviews, and so forth, can be used by practitioners to develop appropriate mathematics programs and to document student progress in the school's curriculum or in a remedial program. When push comes to shove, the following questions should guide our assessment practices:

- Where do students stand in relation to their peers?
- What do students know and what don't they know?
- Why do students perform as they do (i.e., how on earth did the student come up with that answer)?
- Is what I am teaching working?

More than ever, practitioners have the tools to provide answers to these questions. As the ad says, it is time to "Just do it!"

6. Instructional Design in Mathematics for Students with Learning Disabilities

DOUGLAS CARNINE

Miller and Mercer (chapter 4 of this book) described the difficulties students with learning disabilities (LD) have in learning and applying mathematics. As a specific example, Cawley and Miller (1989) reported that 8- and 9-year-olds with learning disabilities performed at about the first-grade level on calculations and application, while 16- and 17-year-old students with learning disabilities scored at about the fifth-grade level.

The potential causes for these difficulties are numerous, including neuropsychological ones (see Rourke & Conway, chapter 3 of this book). Another potential cause of difficulty for students with learning disabilities is the poor fit between the design of mathematics instructional materials and the students' learning characteristics, such as memory skills, strategy acquisition and application, vocabulary, and lanugage coding (Baker, Kameenui, & Simmons, in press). Design features in instructional materials that lead to this mismatch include too rapid a rate of introduction of new concepts, insufficiently supported explanations and activities, and insufficient practice and review (Carnine, Jones, & Dixon, 1994). When concepts are introduced rapidly with minimal explanations and sparse practice and review, students with LD may be overwhelmed by memorization, strategies, vocabulary, and language coding; leading to a steadily growing gap in mathematics attainment between general and special education students.

This gap may be greatest in the area of problem solving, probably because strategies for problem solving entail competence not only in mathematics but also in language. Over the past 25 years, researchers at the University of Oregon have been involved in the development and evaluation of curricular materials for a number of subject areas, including problem solving in mathematics. In the process, we have identified and conducted research on a set of five design principles for improving the quality of instruction for students with learning disabilities: (a) big ideas; (b) conspicuous strategies; (c) efficient use of time; (d) clear, explicit instruction on strategies; and (e) appropriate practice and review.

Although there are many important mathematics objectives for students with LD, this chapter describes and illustrates the five design principles in the context of teaching problem solving. In illustrating the design principles with the most complex content in mathematics—problem solving—this author suggests that the principles would be relevant to simpler content as well. The examples are drawn from the middle grade levels (see Engelmann, Carnine, Kelly, & Engelmann, 1991–1995; Stein, Silbert, & Carnine, in press).

Prior to the discussion of the five principles, the chapter examines the kinds of challenging problem solving that often defeats students with learning disabilities. This examination sets the stage for a discussion of the design principles that might replace that failure with success.

PROBLEM SOLVING, UNDERSTANDING, AND REASONING

A major educational goal for all students, including those with learning disabilities, is better problem-solving performance. Problem solving can be thought of as the application of knowledge to solve a novel problem. Because problem solving covers such a broad range of objectives and activities, it is helpful to work from a sample problem, such as the following:

> Your fifth-grade classroom is going to be in charge of ordering milk for the school. Students can choose either white milk or chocolate milk. Make an accurate estimate of how much white milk and chocolate milk to order each day for the entire school.

The kind of understanding and reasoning involved in the milk-ordering problem implies that students who solve it have acquired relevant knowledge and, depending on the task, can use the knowledge appropriately. More specifically, for a given concept, students need to not only understand what the concept means but also know how to apply it and when to apply it. These ways of using knowledge can be illustrated with proportions, a "big idea" central to solving the milk-ordering problem; students set up the proportion of white milk to chocolate milk for the fifth-grade class and for the entire school.

	Fifth-grade class	Entire school
White milk	_____	_____
Chocolate milk	_____	_____

One way to help students understand the *what* of a new concept is to relate it to a familiar concept. A familiar concept that can serve as a basis for understanding proportions is division. Problem A below is stated and worked as a division problem.

A. A truck holds 8,400 pounds of coal. It delivers the same amount of coal at seven different places. How much coal does each place get?

The division problem indicates that the total weight of 7 equal-sized groups is 8,400 pounds. The solution gives the weight for one group: 8,400 ÷ 7 = 1,200.

Problem B, which is mathematically equivalent to Problem A, is stated and worked as a proportion problem.

B. The truck delivers 1/7 of its load of coal to a school. The truck carries 8,400 pounds of coal. How much coal will be delivered to the school?

A proportion can explicitly express this relationship:

$$\frac{\text{School coal}}{\text{Total coal}} \quad \frac{1}{7} = \frac{\boxed{}}{8400}$$

One out of every seven pounds of coal is delivered to the school. The denominators in the proportion problem express that the seven groups weigh 8,400 pounds; the numerator of 1 indicates that the numerator of the equivalent fraction will give the weight for one group.

Problem B not only develops the *what* (i.e., the proposition described a relationship between school coal and total coal) for proportions, but also illustrates a set of steps for *how* (i.e., write the units, insert the numbers that are known, and solve) students can work a proportion problem. The students set up the units for the proportion (pounds of coal for the school and pounds of total coal) and the proportion itself, placing the numbers in the appropriate place in the proportion. The students then compute the answer—1,200 pounds of coal.

Knowing the *what* and the *how* for a big idea such as proportions is not sufficient for solving problems such as the one for milk ordering, however. Students must also learn about the *when* if they are to successfully solve problems such as A and B in Figure 6.1. Problem A is a positive example of when to use a basic proportion, because applying that strategy for Problem A produces a correct answer. Students who do not adequately comprehend the *when* for using a basic proportion will list 8,400 as the total number of cartons for Problem B, just as they would do for Problem A. The 8,400 is written in the denominator of Problem B across from total cartons, when in fact the 8,400 is for cartons of *apple* juice. Thus, the answer of 2,400 cartons of grape juice is incorrect for Problem B.

More complete knowledge about the *when* of proportions would alert students to the fact that there is not a place to list the 8,400 cartons of apple juice, for the only units in the proportion are for grape juice cartons and total cartons. Students who understand the *when* for using a proportion would

---

Problem A: A truck delivers cartons of juice to a store. 2/7 of the juice is grape. The truck has 8400 cartons of juice. How many cartons of juice will the truck deliver?

	Fraction	Juice Cartons
$\dfrac{\text{Grape}}{\text{Total}}$	$\dfrac{2}{7}$	$=$	$\dfrac{\boxed{2400}}{8400}$

Problem B: A truck delivers cartoons of grape and apple juice to a store. 2/7 of the juice is grape. The truck will deliver 8400 cartons of apple juice. How many cartons of grape juice will the truck deliver?

	Fraction	Juice Cartons
$\dfrac{\text{Grape}}{\text{Total}}$	$\dfrac{2}{7}$	$=$	$\dfrac{\boxed{2400}}{8400}$

Figure 6.1. Positive and negative examples of when to use a basic proportion.

Step 1:	Ratio	Juice Cartons
The students set up the ratio table and fill in the numbers given in the problem.	Grape 2	☐
	Apple ☐	8400
	Total 7	☐
Step 2: The students use their knowledge of addition/subtraction number families to come up with the missing number in the ratio column: $$7 - 2 = \boxed{5}$$	Grape 2 Apple $\boxed{5}$ Total 7	☐ 8400 ☐
Step 3: The students write and solve the proportion to determine the number of cartons of grape juice: $$\frac{2}{5} = \frac{\boxed{3360}}{8400}$$	Grape 2 Apple $\boxed{5}$ Total 7	$\boxed{3360}$ 8400 ☐

Figure 6.2. Proportion and data analysis applied in solving a complex problem.

not inappropriately use a basic proportion with Problem B. Problem B is a negative example of when to use a basic proportion, because applying that strategy in Problem B produces an incorrect answer.

Although a basic proportion cannot be applied with the numbers given in Problem B, the combination of a basic proportion and data analysis allows students to solve for the number of grape juice cartons. The application of this combination is illustrated in Figure 6.2. The combination of basic proportions and data analysis illustrated in Figure 6.2 incorporates the *what*, the *how*, and the *when* of basic proportions. This combination, along with an understanding of probability and estimation, would allow students to solve the milk-ordering problem described near the beginning of this chapter.

First, the fifth-grade students might assume that the preference for types of milk in their class represents the probability of the whole school's preference. On the basis of that assumption, the students could determine a proportion of chocolate milk-to-white milk for their class, find the total enrollment of the school, map all this information in a proportion table, and calculate a solution to the problem. These steps are illustrated in Figure 6.3.

Step 1: The students conduct a survey in their class to determine the numbers for white and chocolate milk. The students also find out from the office the total enrollment for the school.	There are 32 students in the class: 22 prefer chocolate milk and the rest prefer white. There are 479 students in the school.
Step 2: The students organize their data in a ratio table.	**Fifth-Grade Class** / **Entire School** Chocolate 22 ☐ White 10 ☐ Total 32 479
Step 3: The students solve a proportion: Determine the estimated number of chocolate milk cartons for the entire school. $$\frac{22}{32} = \frac{329}{479}$$	Chocolate 22 329 White 10 ☐ Total 32 479
Step 4: The students determine the last missing number using their knowledge of addition/ subtraction number families: $$479 - 329 = 150$$	Chocolate 22 329 White 10 150 Total 32 479

Figure 6.3. Proportion, data analysis, and probability applied in solving a complex problem.

There are many extensions of the milk-ordering problem that students also would be able to solve. Actual attendance could be used instead of total enrollment, which would cut down on milk ordered (and wasted). The students could also determine how well their preferences for white or chocolate milk represent those of all the fifth-grade classes. They could gather data from the other fifth-grade classes and compare it with their own data. Finally, they could evaluate the accuracy of their estimate by comparing it with the number for the type of milk the students actually selected.

The milk-ordering problem and these extensions meet many goals of the standards of the National Council of Teachers of Mathematics (1989). As students discuss their options for selecting a sample group (e.g., a single fifth-grade class, all the fifth-grade classes, one class from each grade level, etc.), they are working together to enhance their understanding of mathematics. As they weigh the relative merits of using total enrollment versus actual attendance, they are engaging in conjecture and invention. As they link their understandings of the various big ideas of proportion, data analysis, and probability, they are learning to connect mathematical ideas, solve problems, and apply mathematics broadly. The crucial question is: How can students with learning disabilities not only learn the basics in mathematics, but also successfully engage in these important and challenging activities as described in the standards of the National Council of Teachers of Mathematics?

PRINCIPLES OF EFFECTIVE INSTRUCTIONAL DESIGN

Teach Big Ideas

Solving the milk-ordering problem is difficult for most students, not just those with learning disabilities. Certain design principles could benefit all students, one of which is to organize content around "big ideas," which is the first of the five design principles. Big ideas represent central ideas within a discipline that will make learning "subordinate" concepts easier and more meaningful. They have rich explanatory and predictive power, as students can use them in solving many different problems that on the surface appear to be unrelated (Carnine & Kameenui, 1992; Carnine & Shinn, 1994). And, perhaps foremost, big ideas apply to many common, everyday contexts and situations.

Unfortunately, this first design principle has been empirically investigated in science but not in mathematics. Using a big idea in many different science situations has improved problem solving and deepened understanding (e.g., Woodward, 1994). The previous section of this chapter illustrated how students can orchestrate several big ideas—proportion, data analysis, and probability—to solve a complex problem.

Another example of a big idea is volume. Students usually learn seven distinct formulas for volume. A big idea reduces the number of formulas students must learn to three slight variations of a single formula—the area of the base times the height (B • h). Expressing volume as area of the base times a multiple of the height enhances understanding while simultaneously reducing the quantity of content to be taught.

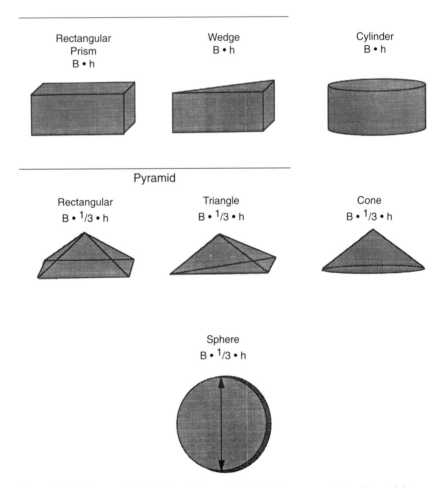

Figure 6.4. Volume as the "big idea" of area of the base times a multiple of the height.

Seven figures are displayed in Figure 6.4. For the three figures in which the sides go straight up—rectangular prism (box), wedge, cylinder—the volume is the area of the base times the height (B • h). For figures that come to a point (pyramid with a rectangular base, pyramid with a triangular base, and a cone), the volume is not the area of the base times the height but, rather, the area of the base times 1/3 of the height (B • 1/3 • h). The sphere is a special case—the area of the base times 2/3 of the height (B • 2/3 • h)—where the base is the area of a circle that passes through the center of the sphere, and the height is the diameter. This analysis of big ideas fosters understanding of the key concept that volume is a function of base times height. As Gelman (1986) stated, "a focus on different algorithmic instantiations of a set of principles helps teach children that procedures that seem very different on the surface can share the same mathematical underpinning and, hence, root meanings" (p. 350).

Instruction on big ideas can benefit general education students as well as students with LD. However, the characteristics of students with learning disabilities and research on how they learn imply that there are more effective and less effective ways of teaching big ideas. Next discussed are four design principles that are particularly relevant for students with learning disabilities. A number of studies have investigated these principles in various content areas; however, many of the investigations did not include students with learning disabilities as participants, making the research base less robust than is desirable.

Teach Conspicuous Strategies

The second principle is to teach conspicuous strategies. A strategy is a series of steps that students follow to achieve some goal. Such steps are an approximation of the steps experts follow covertly (and, perhaps, unconsciously) while working toward similar goals. In instruction, such steps are initially made overt and explicit for students. Eventually, as students master a strategy, the steps become more covert, as for experts. Any routine that leads to both the acquisition and utilization of knowledge can be considered a strategy (Prawat, 1989). Students who are able to solve the milk-ordering problem appear to possess strategies that facilitate using knowledge in a variety of ways—the *what, how,* and *when.*

A strategy may be so specific and narrow in application that it is little more than a rote sequence for solving a particular problem or a very small set of highly similar problems. For example, students in one treatment of a problem-solving study learned a rote formula (+1 −2 −2), which, predictably, did not transfer well in solving new problems (McDaniel & Schlager, 1990). At the other extreme, a strategy may be so general that it is little more than a broad set of guidelines or heuristics that, while better than nothing, do not dependably lead most students to solutions for most problems. For instance, a strategy such as "draw a picture" is probably far too general for reliably leading a majority of students to reasonable solutions to the juice-ordering problem. Finally, some students develop strategies that are neither too narrow nor too broad. These strategies are "just right."

One challenge of instruction—perhaps the major challenge—is to develop "just right" strategies for interventions with those students who do not develop them on their own, including, but not limited to, at-risk students and students with learning disabilities.

On the basis of an exhaustive review of research, Prawat (1989) recommended that efficient strategy interventions be intermediate in generality. That is, efficient strategies fall somewhere between the extremes of narrow application (but, presumably, relatively easy to teach) and broad (but not necessarily reliable or easy to teach). This suggests that the principal feature of a "good strategy" is that it adheres to the Law of Parsimony as it applies to evaluating competing theories: The best theory explains the most in the simplest way (Mouly, 1978). As applied specifically to evaluating strategies, the Law of Parsimony might read: The greatest number of students from a targeted student population can successfully solve the greatest number of problems or complete the broadest range of tasks by applying the most economical strategies. Various forms of strategy instruction have proved to be effective with students with LD in both mathematics (Montague & Bos, 1986) and reading (Paris, Lipson, & Wixson, 1983).

Use Time Efficiently

The third principle is to use time efficiently. This principle addresses a perennial problem in teaching at-risk students and students with learning disabilities: teaching all they need to know (in both quantitative and qualitative terms) without "losing" them by trying to do too much, too quickly. Catching them up with their peers, as efficiently as possible within realistic time constraints, should not overwhelm the students.

Abandon Low-Priority Objectives and Focus on Big Ideas. Understanding new information and how it interrelates with previously acquired knowledge may be inherently complex. Rather than trying to strip such knowledge of its complexity (and therefore, quite probably, of its meaningfulness), the information can be analyzed for big ideas. In most cases, mathematics programs attempt to cover exhaustive lists of learning objectives, with little or no attempt to prioritize those objectives on the basis of their relative importance later. For example, basal mathematics texts typically teach from 8 to 11 different problem-solving strategies. Each strategy is taught for only one lesson and receives minimal review. If instruction focuses on all the reasonably important big ideas, then not only is understanding enhanced, but new material can be introduced less frequently, which reduces the likelihood of students' being overwhelmed.

"Ease Into" Complex Strategies. A variation of introducing too much new material is the traditional practice of teaching complete complex strategies in toto within a single instructional encounter. One example would be introducing the complex strategy for the milk-ordering problem in a single lesson. Students would be expected to integrate proportion, data analysis, and probability all at once. This practice would likely overwhelm most students with learning disabilities.

There is no reason why instruction on complex strategies comprising many component strategies cannot be spread out over a few days' time—a method of simplifying the communication of complexity without sacrificing crucial inherent complexities (Carnine, 1980; Kameenui & Carnine, 1986; Kameenui, Carnine, Darch, & Stein, 1986; Paine, Carnine, White, & Walters, 1982). For example, instead of having to analyze the milk-ordering problem and create a table for the values, students would be given a table like the one in Figure 6.5 and use it to solve for the number of cartons of white and chocolate milk for the entire school. Next, the students would not be given the table, but would be taught to construct a table before solving a problem.

	CLASS	SCHOOL	
WHITE MILK	22		
CHOC. MILK	15		
TOTAL	37	680	

Figure 6.5. Proportion/data analysis table to help students "ease into" a complex strategy.

Use a Strand Organization for Lessons. The relatively small numbers of big ideas that are selected for instruction do not have to be presented in their entirety in a single lesson. Each big idea and its component concepts can be taught as a strand. A strand is a portion of several consecutive lessons that is devoted to a particular big idea. Rather than organizing an entire lesson around a single topic, as is done in traditional basal programs, lessons can be designed around strands; for example, in a lesson, each 5- to 10-minute portion from a strand can address a different big idea.

There are several reasons for organizing instructional material around strands. First, students are more easily engaged with the variety that strands offer. For example, 30 minutes on renaming (borrowing), day in and day out, would become quite tedious. In contrast, a lesson consisting of 8 minutes on renaming followed by 6 minutes on estimation, 3 minutes on facts, and 15 minutes on verbal problems would be more likely to keep students engaged. With varying content from each strand, students are more attentive and have time to practice and review several concepts. Working 60 or 70 problems consisting of a mix of renaming, estimation, facts, and word problems is reasonable; working 60 or 70 borrowing problems in a lesson can be tedious for many students with learning disabilities and is inappropriate.

Second, strands make the sequencing of component concepts more manageable. Big ideas in mathematics contain many concepts. Arranging these component concepts in a scope and sequence such that they are taught prior to their integration is made easier when several of them can appear in one lesson. This particular advantage of a strand organization was illustrated earlier with "easing into" complex strategies (see Figure 6.5).

Third, lessons organized around strands make cumulative introduction feasible. In cumulative introduction, after a concept is introduced, it is systematically reviewed and integrated with other related concepts. Cumulative introduction, as an alternative to the traditional spiral introduction, has three important advantages: (a) Component concepts of complex strategies can be introduced early, (b) practice can be provided on both new and previously introduced concepts until responses are accurate and rapid, and (c) distributed practice on some concepts can occur every day. For example, complex proportion/data analysis problems may need to be practiced for several lessons until students can work this type with relative ease (massed practice). Then a mixture of simple and complex proportion problems might be presented, to ensure that students know when to apply each strategy. Finally, only one or two problems of both or either type could be presented daily (distributed practice). Distributed practice is easy to schedule when each lesson is designed to incorporate portions of several strands.

Use Manipulatives in a Time-Efficient Manner. Evans (1990) compared instruction that began with manipulatives (followed by work with a meaningful algorithm) and instruction that began with a meaningful algorithm (followed by work with manipulatives). The students in both sequences ended up with comparable (but relatively low) understanding and proficiency. However, beginning with manipulatives was less efficient; students in the sequence that began with manipulatives took, on the average, 75 more minutes to learn to work with manipulatives and the algorithm. That study suggests that sequencing manipulatives as a follow-up activity is more efficient than using them as an introductory activity.

Many educators believe that support for manipulatives is so unequivocal that efficiency concerns should be disregarded. However, several prominent researchers have urged caution about the uncritical use of manipulatives. "Perhaps manipulatives should carry the following warning label: The Secretary of Education [or other appropriate authority] has not determined that using manipulatives is either a sufficient or a necessary condition for meaningful learning" (Baroody, 1989, p. 4). Resnick and Omanson (1987) go further, questioning whether blocks or manipulatives play a crucial role in learning regrouping in subtraction, and suggest that "perhaps any discussion of quantities manipulated in written arithmetic, without any reference to the block analog, could be just as successful in teaching the principles that underlie written instruction" (p. 90). Moser (1980) and Bana and Nelson (1977) found that the use of manipulatives with young children resulted in poorer problem solving. As Heibert (1984) observed, "It is not just the use of concrete materials that improves mathematical understanding, but rather the explicit construction of links between understood actions on the objects and related symbol procedures" (p. 509).

Clearly Communicate Strategies in an Explicit Manner

The fourth principle is to clearly communicate strategies in an explicit manner. New concepts and strategies should be explained in clear, concise, accurate, and comprehensible language. Here are specific guidelines for facilitating clarity:

Make Strategies Explicit and Clear. Experts implement covert strategies as their means of acquiring and utilizing knowledge. The whole purpose of developing instructional strategies is to explicate expert strategies so that they become clear to nonexpert learners. The research support for such explicit instruction is quite strong across domains (e.g., in mathematics: Carnine & Stein, 1981; Charles, 1980; Gleason, Carnine, & Boriero, 1990; Leinhardt, 1987; Resnick, Cauzinille-Marmeche, & Mathieu, 1987; Resnick & Omanson, 1987; Stigler & Baranes, 1988; in health science: Woodward, Carnine, & Gersten, 1988; in language arts: Bereiter & Scardamalia, 1987; Graham & Harris, 1988; Isaacson, 1987; Thomas, Englert, & Gregg, 1987). Moreover, explicit instruction has been shown to be effective for teaching at-risk students (Carnine, Engelmann, Hofmeister, & Kelly, 1987; Egan & Greeno, 1973; Fielding, Kameenui, & Gersten, 1983; Kameenui et al., 1986; Kelly, Carnine, Gersten, & Grossen, 1986).

Following is an example of providing clear, explicit instruction for introducing proportion word problems. Students would already have learned the *what* and the *how* for proportions. Consider the following problem: "Juan's sister tells him that there will be a 5-second gap between seeing lightning and hearing thunder for each 1 mile of distance. Juan counts 20 seconds between the lightning and thunder. Juan is to figure out how many miles."

Here is what the teacher might say and do:

"What does the problem ask you to figure out?" (Teacher pauses before calling on individual students or groups of students if they are working cooperatively.)

"Yes, the problem asks about how many miles away the lightning is. The gap between seeing and hearing the lightening is 20 seconds. For every 5 seconds, the gap is 1 mile. That's a ratio of 5 to 1, so we write a 5 for seconds and a 1 for miles."

(Teacher writes on board:

$$\frac{\text{seconds}}{\text{miles}} \quad = \quad \frac{5}{1} \text{)}$$

"The problem gives us a number for total seconds. Find the sentence that tells about how many seconds in all." (Teacher waits, then solicits responses.)

(Teacher writes 20 for seconds and $\square$ for total miles:

$$\frac{\text{seconds}}{\text{miles}} \quad \frac{5}{1} \quad = \quad \frac{20}{\square} \text{)}$$

"Now we can figure out how many miles for 20 seconds."

The importance of explicit instruction cannot be assessed without reference to other design principles, such as the quality of a conspicuous strategy (the second principle). Explicit teaching of a rote strategy can effectively produce rote learning, which is illustrated in a study by McDaniel and Schlager (1990) involving water-jar problem solving (see Note). Students in the explicit teaching treatment learned a rote formula (+1 −1 −2), which did not transfer well in solving other water-jar problems. Quite often in research, explicit instruction is applied to content that has not been designed to be generalizable, such as McDaniel and Schlager's formula of +1 −1 −2. These interventions should be called *explicit rote teaching.* The flaw, however, is in the design of the strategy, not the explicit teaching.

Accommodate Differences in Prior Knowledge. The phrase *prior knowledge* is used ambiguously in the literature. Sometimes *prerequisite knowledge, background knowledge,* and *prior knowledge* are used interchangeably. In other instances, these terms are differentiated from one another. We will use *prior knowledge* to refer to the following varying conditions of knowledge: verbal background knowledge and component concept knowledge.

Verbal background knowledge first became prominent with respect to reading comprehension. For instance, a student familiar with sailing and sailboats is much more likely to comprehend a story about sailing than is a student without such background familiarity. Such knowledge is potentially important in mathematics, principally in conjunction with solving word problems. A student who does not know that a truck is a type of vehicle is clearly preempted from solving classification problems involving cars, trucks, and vehicles.

The component concept knowledge required for some strategies is extensive. Most students with LD cannot learn in one class period all the concepts that come into play in a complex strategy. In these cases, the components must be established as prior knowledge and then combined. For example, there are several levels of difficulty in computing answers to proportion problems. In the simplest problems, the first denominator is a factor of the second denominator; in this example, 5 is a factor of 10:

$$\frac{3}{5} = \frac{\square}{10}$$

A slightly more difficult type involves larger factors:

$$\frac{12}{18} = \frac{\square}{54}$$

A third type has fractions in which the first denominator is not a factor of the second denominator; in this example, 5 is not a factor of 12:

$$\frac{3}{5} = \frac{\square}{12}$$

Not all types of proportion computation problems need to be taught before any proportion word problems are introduced. However, instruction on more difficult proportion computation should come before word problems that require difficult computation.

The introduction of any new strategy is likely to require quick and easy access to verbal knowledge or component concept knowledge. Instruction should accommodate that requirement by testing either to ensure that students possess prerequisites, or to determine what prerequisites need to be taught or retaught before proceeding to the new strategy.

A variety of studies indicate that students benefit from learning the components of strategies as a prerequisite to mastering those strategies themselves (Barron, Bransford, Kulewicz, & Hasselbring, 1989; Darch, Carnine, & Gersten, 1984; Leinhardt, 1987; Nickerson, 1985).

Provide Scaffolded Transition to Self-Directed Learning. In addition to making initial instruction as explicit and as clear as possible, in the senses

---

Table 6.1
Scaffolded Teacher Wording for Solving a Proportion Problem

A security guard walks 4,800 feet during the 8 hours he works. He walks about the same number of feet each hour. How many feet does walk by the end of the fourth hour?

Teacher says:	Students write:
• Find the question you're going to answer.	
• Write the units and the numbers for the ratio. (Check.)	$\dfrac{\text{feet}}{\text{hours}} \quad \dfrac{4800}{8}$
• Find the other number that is given and use it to write the other ratio. (Check.)	$\dfrac{\text{feet}}{\text{hours}} \quad \dfrac{4800}{8} = \dfrac{\square}{4}$
• Solve for the unknown and write the complete answer. (Check.)	$\dfrac{4800}{8} = \dfrac{2400}{4}$

---

described earlier, instruction can be scaffolded (also referred to in the literature as *prompting and guiding;* Tharp & Gallimore, 1988). Imagine that you wanted to help a toddler play on a slide. It is not likely that you would just show that child how to climb the ladder and slide down, and then say, "You're on your own now." More likely, you would help that child as much as necessary, gradually withdrawing your help as he or she became more confident and competent. Scaffolding is that same type of assistance given to students in academic areas between the introduction of new knowledge and the eventual self-directed application of that knowledge.

There is a variety of means by which instruction can be scaffolded between initial instruction and self-directed learning, including the creation of scaffolded worksheets illustrated in the proportion/data analysis table in Figure 6.5. Scaffolding can also come from peer coaching, cooperative learning, and self-monitoring. However, teacher directions and questions, as illustrated in Table 6.1, are usually the most important source of scaffolding. The teacher tells the students what to do, step by step, and checks their work after each step.

In response to the finding that students with learning disabilities and other at-risk students often falter as self-directed learners, many theorists from a variety of traditions recommend scaffolding as a gradual transition from teacher direction to self-direction (Engelmann & Carnine, 1991; McDonnell & Ferguson, 1989; Miller & Test, 1989; Pellegrino & Goldman, 1987; Resnick, 1988; Rosenshine & Stevens, 1984; Scardamalia & Bereiter, 1986; Trafton, 1984).

Use Consistent Feedback. The research evidence supporting corrective feedback is extremely strong. Many studies across content areas and type of students have focused directly on feedback effects (e.g., Armour-Thomas, White, & Boehm, 1987; Barbetta, Heward, Bradley, & Miller, 1994; Lhyle & Kulhavy, 1987; Prueher, 1987; Schimmel, 1983; Siegel & Crawford, 1983). In others, corrective feedback has been studied as an important element within an instructional system (e.g., Fuchs, Fuchs, & Hamlett, 1989; Guskey & Gates, 1986; Richman & Brown, 1986).

Some theorists argue that the information given during corrective feedback in the scaffolded instruction phase should differ from that given during initial instruction, based upon the assumption that the explanation given during initial instruction must have been inadequate—and, hence, student errors occurred (e.g., Guskey & Gates, 1986). For example, a student wrote the following in working the problem in Table 6.1 about the security guard:

$$\frac{\text{feet}}{\text{hours}} \quad \frac{4800}{8} \quad = \quad \frac{4}{\square}$$

The student made an error in placing the number and symbol for the unknown. The teacher corrects by presenting an alternative strategy: The student is told to divide the total feet by the number of hours, $4,800 \div 8 = 600$, and then to multiply the answer by 4. The introduction of this second algorithm to the student who is already confused might actually increase his or her confusion.

An alternative is to reteach the original strategy. Reteaching the original strategy is preferred if there are efficacy data supporting the original strategy, such as that produced in learner-verification studies (Collins & Carnine, 1988). In learner verification, student and teacher difficulties with a strategy are noted and the strategy is revised and then used again. This process is repeated until no substantive difficulties are encountered. Once a "best" approach to initial instruction has been refined and verified, the same strategy should be used for initial instruction and for corrective feedback. If student errors persist, the way in which the instruction is being delivered should be evaluated. If the way in which the instruction is being delivered is appropriate and errors persist, then an alternative approach should be tried.

Provide Practice and Review to Facilitate Retention

The fifth principle is to provide practice and review to facilitate retention. One goal of a mathematics instructional program is for students to remember and apply increasingly complex concepts and strategies. To develop retention, students must receive carefully planned practice opportunities, spaced over a period of time. Following are several guidelines.

1. *Facilitate automaticity, when appropriate.* Adequate practice should be provided to enable students to reach a point of performance that allows them to readily apply a concept as a component of more complex strategies. For example, when learning how to work simple proportion word problems, students first observe demonstrations and then receive thorough scaffolded instruction. This massed practice continues until the students are facile at setting up the proportion and working the problem.

Practice then switches from massed to distributed, or spaced practice. This can occur in one of two ways: (a) Problems can appear every second, third, or fourth day or (b) the strategy can be subsumed within a more complex strategy. For example, distributed practice in the strategy for simple proportion word problems could occur every third or fourth lesson until the introduction of the proportion/data analysis strategy, which subsumes the simpler proportion strategy. Incorporating a simpler strategy into a more complex one does not mean that the simpler strategy should not continue to be applied, a point that is discussed next.

2. *Provide integration opportunities.* As noted above, the amount of practice on problems of a newly introduced type should be great enough that the student can work the problems with relative ease. After initial instruction, problems of that type should be mixed in with previously introduced problems of a similar nature (Kelly, Gersten, & Carnine, 1990). This mix provides the student with practice not only in the *how* of a strategy but also the *when* of a strategy. For example, after more complex proportion/data analysis problems are initially practiced, a mix of these problems and simple proportion problems should be presented. Problems A and B in Table 6.2 illustrate a mix of problem types.

CONCLUSIONS

Building instructional programs and carrying out instruction based on the five design principles discussed in this article should enable students with learning disabilities both to acquire basic skills and to solve challenging problems in mathematics. Such successes will result in part because of the accommodation of the characteristics of students with learning disabilities by the design principles. However, there is relatively little research on applying

---

Table 6.2
Integration of More Complex and Simple Problems

A. The average class drinks 14 cartons of white milk and 16 cartons of chocolate milk each day. The whole school drank 240 cartons of chocolate milk. About how many cartons of white milk did the students in the school drink?

B. The average class drinks 14 cartons of white milk and 18 cartons of chocolate milk each day. There are 240 students in the school. About how many cartons of white milk do the students in the school drink?

	Class	School
White	14	210
Chocolate	16	240

$$\frac{14}{16} = \frac{\boxed{210}}{240}$$

	Class	School
White	14	—
Chocolate	16	—
Total	30	240

$$\frac{14}{30} = \frac{\boxed{112}}{240}$$

---

the principles with secondary students with learning disabilities (e.g., Moore & Carnine, 1989). (See Jones, Wilson, & Bhojwani, chapter 8 of this book, for a discussion of instruction for secondary students.) Also, learning to solve problems will take far more time for students with severe learning disabilities. Success is more likely if students receive careful instruction with well-designed instructional materials throughout the elementary grades (Gersten & Carnine, 1982).

Secondary students who do not receive such instruction in the elementary grades can benefit from problem-solving instruction even if they lack computational skills (e.g., they could use calculators). These types of adjustments are relatively easy to make in the context of a school program designed to effectively and efficiently teach challenging content to students with learning disabilities. Unfortunately, creating such schools is very difficult.

One step toward the creation of such schools is the empirical validation of design principles such as the five described in this article and the acceptance of the principles by faculty in preservice institutions, by directors of curriculum departments in school districts, by publishers of mathematics textbooks and other educational tools, and by officials in government agencies. The path to validation and then to acceptance is largely uncharted (see Carnine, 1995, for some preliminary suggestions).

AUTHOR'S NOTE

This research was supported in part by Grant No. H180M10006-94 from the Office of Special Education Programs. The views expressed within this chapter are not necessarily those of the U.S. Department of Education.

NOTE

The object is to figure out a sequence of filling and emptying operations that will result in the goal amount residing in one of the uncalibrated jars. For example, the subject is told that Jar A can hold 7 units, Jar B can hold 24 units, and Jar C can hold 98 units. The goal is to obtain 60 units of water in Jar C. To solve this problem, one must first fill Jar C from the sink, then, using Jar B, take out 24 units from Jar C, leaving 74 units. Next, Jar A is used to take 7 units from Jar C, leaving 67 units in Jar C. Finally, Jar A is again used to take 7 units from Jar C, resulting in the goal of 60 units. Thus, the solution for this problem is found via the equation 1C –1B –2A.

7. Mathematics Instruction for Elementary Students with Learning Disabilities

CAROL A. THORNTON, CYNTHIA W. LANGRALL, AND

GRAHAM A. JONES

The curriculum, assessment, and professional teaching *Standards* of the National Council of Teachers of Mathematics (NCTM; 1989, 1991, 1995) call for strategic shifts in mathematics instruction for all students. In essence, these shifts involve a movement toward higher level mathematical reasoning and problem solving, and involve rethinking long-established beliefs about teaching, learning, and curricular practices. Common practice in both general and special education classrooms, however, still reflects a narrow emphasis on computation. This focus is also mirrored in diagnostic teaching and evaluation thrusts (Heshusius, 1991). Not only are such perspectives on instruction and assessment incompatible with the vision of the Standards, but they are also contrary to the findings of recent research about mathematics teaching and learning (e.g., Carpenter, Fennema, Peterson, Chiang, & Loef, 1989; Englert, Tarrant, & Mariage, 1992; Resnick, 1987, 1989; Thornton & Bley, 1994).

The National Council of Teachers of Mathematics (1989) proposed five goals for rethinking mathematics teaching and learning. The council held that students should (a) learn to value mathematics, (b) become confident in their ability to do mathematics, (c) become mathematical problem solvers, (d) learn to communicate mathematically, and (e) learn to reason mathematically. To accomplish these goals, the Council advocates, teachers should decrease their emphasis on complex paper-and-pencil computation, rote memorization of rules and formulas, written practice, "one answer, one method," and teaching by telling.

These recommendations for school mathematics are grounded in constructivist theory (e.g., Cobb & Bauersfeld, 1995; Noddings, 1990) and stem from a broad research base in mathematics education (Grouws, 1992). Little of this research, however, has focused specifically on mathematics instruction for students with learning disabilities (LD). Furthermore, few studies involving students with learning disabilities have focused on higher level mathematical thinking and problem solving (Parmar & Cawley, chapter 11 of this book; Marshall, 1988; Mastropieri, Scruggs, & Shiah, 1991).

A major goal of this chapter is to describe four major themes related to higher level thinking and problem solving that have emerged from recent mathematics studies involving students with learning disabilities. These themes are as follows:

• Providing a broad and balanced mathematics curriculum;
• Engaging students in rich, meaningful problem tasks;
• Accommodating the diverse ways in which children learn; and
• Encouraging students to discuss and justify their problem-solving strategies and solutions.

In essence, these themes embrace the philosophy that students with learning disabilities benefit from rich, challenging programs that promote mathematical thinking. Each of these themes will be illustrated by case study examples drawn from one or more of four research reports. Prior to considering the case studies, we present an overview of each in the next section. These overviews are intended to provide a context for describing and discussing the case studies.

STUDIES INVOLVING STUDENTS WITH LD

Study 1: Children's Understanding of Multidigit Numbers

This study by Jones et al. (1996) validated a framework that described and predicted various levels of young children's thinking related to multidigit number sense. The framework itself was used to generate and evaluate two different versions of an instructional program emphasizing number sense in general education classrooms that included students with identified learning disabilities.

The instructional approach based on the framework, grounded in social constructivism (e.g., Cobb & Bauersfeld, 1995), took the position that students' opportunities to construct mathematical knowledge arise from attempts to resolve conflicting points of view in a group, from attempts to reconstruct and verbalize a mathematical idea or solution and, more generally, from attempts to reach consensus with others. The instructional program was consistent with the recommendations of Englert et al. (1992) and Heshusius (1991), who suggested that students with disabilities should be challenged with meaningful problem tasks that promote multiple solutions and strategies.

Throughout the study, all students were encouraged and given time to work collaboratively to solve problems at whatever levels they could attain. A further expectation was that all students would share and justify their thinking in different ways. Differences in the understandings demonstrated by children in the two instructional groups were attributed largely to the quality of the problem-solving experiences and the level of student interactions.

Study 2: The Use of Reflective Analyses on the Instructional Practices of Prospective Elementary Teachers

This research by Langrall, Thornton, Jones, and Malone (1996) used a case study approach to investigate the effects of reflection on prospective elementary teachers' instructional practices in mathematics. Recommendations from the NCTM (1989, 1991) Standards documents provided the basis for the teachers' reflections on their instructional experiences in elementary classrooms.

Six prospective teachers participated in a series of 11 instructional classroom experiences. In the first and last of these sessions, the teachers instructed small groups of students, including children with learning disabilities. Video analysis and transcripts of these lessons were supplemented with other data, including semistructured interviews, stimulated-recall sessions, written documents submitted by the teachers, and researcher field notes. These data sources became the context for six case studies that documented the changes in instructional strategies adopted by the teachers.

This intervention resulted in powerful changes in the instructional practices of these prospective teachers, including changes in the way they related to students with learning disabilities. Major changes included (a) greater use of problem-driven tasks and open-ended questions, (b) increased expectations for student reasoning and multiple-solution strategies, (c) greater emphasis on student dialogue and collaboration; and (d) less teacher-directed instruction.

Although the focus of this study was on prospective teachers, the research also captured rich vignettes of student interactions with each other and with their teacher.

Study 3: Supporting Middle School Students with LD in the Mainstream Mathematics Classroom

The major goal of this research project by Borasi, Packman, and Woodward (1991) was to institute a comprehensive professional development program that would encourage and support middle school mathematics teachers in rethinking their teaching goals and practices so as to better respond to the learning needs of *all* their students, with special attention to those with learning disabilities. The project was developed by an interdisciplinary team comprising mathematics educators, an expert in learning disabilities, and both mathematics and special education teachers. Its theoretical framework was characterized by a constructivist perspective on knowledge and learning, an information-processing model of learning disabilities, and an "inquiry approach" to teaching mathematics.

At the core of this project was the development of three units intended to illustrate an "in-action" inquiry approach in middle school classrooms that included students with learning disabilities. Extensive case study data on the students with LD were collected during the implementation of these units in different instructional settings (a private school for students with learning disabilities, and both urban and suburban public school general education classrooms). The results suggest that an inquiry approach, complemented by appropriate instructional modifications and adaptations, can help mathematics teachers meet the challenges that accompany a diverse student population. Furthermore, in mathematics classrooms informed by such an approach, students' learning differences could be capitalized upon and turned into assets in the learning setting.

Study 4: Mathematical Problem-Solving Processes of Primary-Grade Students Identified with LD

Behrend's (1994) study examined the problem-solving processes of five second- and third-grade students identified as having learning disabilities. In accordance with the recommendations of their Individualized Education Programs (IEPs), these students received daily mathematics instruction in an LD resource room. Cognitively Guided Instruction (Fennema & Carpenter, 1985) provided the framework for assessing the children's independent and assisted problem-solving abilities. Data were gathered during individual interviews and small-group sessions. During the group sessions, students were presented with word problems, given time to solve them, and encouraged to share their strategies in whole-group discussions. Prompts, from general to more explicit, were given only when needed.

Behrend (1994) found that, given the opportunity, the students in her study were capable of sharing their strategies, listening to other children's strategies, discussing similarities and differences among strategies, justifying their thinking, and helping each other understand word problems. Although students modeled their solutions for each other, she observed that *teacher* modeling of solution strategies was rarely needed and generally did not promote better problem solving among the children.

All five students could solve a variety of problems, including difficult addition, subtraction, multiplication, and division word problems; problems with extraneous numerical information; and multiple-step problems. Although other studies involving children with learning disabilities have found explicit instruction in strategy implementation to be effective, Behrend (1994) found that the students in her study were capable of generating and utilizing their own problem-solving strategies and did not need to be taught specific strategies. Based on this finding, she questioned the need for explicit strategy instruction in mathematics for students with learning disabilities and recommended instructional approaches that utilize students' available problem-solving processes.

CASE ILLUSTRATIONS OF THE FOUR THEMES

Four case illustrations from the studies described above exemplify the themes presented in this chapter. Although some of these cases could illustrate more than one theme, the discussion highlights salient features of each in turn.

Providing a Broad and Balanced Mathematics Curriculum

Overview of Theme. Baroody and Hume (1991) noted that many children who experience learning difficulties in mathematics, including those who have learning disabilities, are "curriculum disabled." For such students, Trafton and Claus (1994) recommended a broader and more balanced curriculum, in contrast to a more traditional curriculum, with its repetitive and unnecessary emphasis on computation.

A broader curriculum can be established by utilizing problem-driven instruction that incorporates a greater emphasis on number sense and estimation, data analysis, spatial sense and geometric thinking, patterns and relationships leading to algebraic understandings, and the supportive use of technology (National Council of Teachers of Mathematics, 1989). Broadening the curriculum to include a variety of mathematical domains does not preclude, but actually encourages, the development of appropriate mathematical skills. In essence, it presents opportunities for different kinds of thinking and success beyond numerical reasoning. Such a curriculum should be reflected in children's IEPs. Although it does not entail disregarding computation in the IEP, a broadened thrust enables students with learning disabilities to use mathematics more flexibly, insightfully, and productively (Bley & Thornton, 1994; Borasi, in press; Englert et al., 1992).

Case Illustration. To exemplify the "broad and balanced curriculum" theme, this section draws on case data from two students with identified learning disabilities. These illustrations include an episode involving a student we will call Jana, focusing on mental math (Jones et al., 1996), and an episode documenting "Terrell's" thinking about a geometry task (Langrall et al., 1996).

On the Wechsler Intelligence Scale for Children–III (WISC-III), 9-year-old Jana scored well below average on the Verbal Comprehension and Mathematics subtests. Because she also had severe receptive language and auditory memory difficulties, she was placed in a self-contained learning disabilities program. Despite her difficulties with word problems, her good visual memory and interest in mathematics led to her being jointly placed in a general classroom for mathematics, where the following episode occurred. The activity had started the moment Mrs. Tate's class entered the room. Each pair of children selected a card that showed an amount of money they could spend. The task was to "purchase" items from the garage sale mural on the wall, spending as much of their money as possible. The children worked in pairs for a short time before Mrs. Tate brought them together to share their thinking. Over the past 2 weeks the garage sale activity had constituted a daily math problem for the children—one that had grown out of earlier problem-solving experiences with addition and money.

> Jana spoke for herself and her partner: "We picked the picture frame for 38¢ and the poster for 15¢—and we just have 7¢ change." When asked to explain how they knew they would have 7¢ change, Jana said, "We just thought about the 100s Chart. We started with 38 and went down to 48, then counted 5 more. So we paid 53¢ — that gives us 7¢ back because we had 60¢ to spend."

This episode illustrates how the 100s Chart enabled Jana to move beyond paper-and-pencil computation. During early instruction with this graphic aid, Jana was encouraged to move a finger along the chart as she counted. She then grew able to visualize the counting-on process just by thinking of the 100s Chart. In this case the chart was an appropriate compensatory tool that enabled Jana to compute two-digit sums mentally.

The second case centers on Terrell, a student with learning disabilities enrolled in a general fifth-grade class. He showed little ability to reason abstractly and had visual-perception problems, but retained information once he had internalized it. The following episode illustrates how Terrell, his learning difficulties notwithstanding, manipulated pattern blocks and drew on his understanding of a high-dive flip to reason about angle measures in a meaningful way.

> Terrell pointed to the three trapezoids he had arranged around a dot on the overhead [see Figure 7.1]. He explained what Duane had told his group about how a "360 flip" off the high dive "goes all the way around." "Here three of these [trapezoids] go all the way around. So we divided 360 by 3 and got 120 for the big angle."

As he observed groups at work, the teacher, Mr. Adams, had not been sure that Terrell understood Duane's explanation of a 360-degree turn, so he was pleased when he heard Terrell rephrase the explanation to Duane and later volunteer to present the group's solution. Mr. Adams's expectation that all group members would be able to present the group's solution set the stage for Terrell to verbalize his solution strategy within his working group. Expectations like this that encompass opportunities for children with learn-

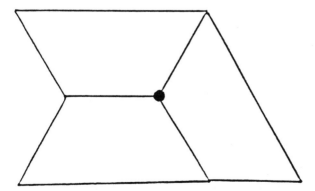

Figure 7.1. Terrell's pattern block illustration.

ing disabilities to articulate their thinking have been shown to help with learning and retention (see Montague, chapter 9 of this book).

Other groups in the class had found different ways of showing that the obtuse angle of the red pattern block was 120 degrees. The initial task had challenged each group of four students to determine the measures of each of the angles of the pattern blocks. As part of the ensuing discussion, a summary chart was made to organize the class's findings.

These case illustrations document how two students with learning disabilities were successful in mathematics programs that emphasized a broad, balanced curriculum. When instruction is consistent with the tenets of such a curriculum and different approaches are valued, it is possible for children to achieve success within their specific limitations (Bulgren & Montague, 1989; Cawley, Fitzmaurice-Hayes, & Shaw, 1988; Ginsburg, chapter 2 of this book).

Engaging Students in Rich, Meaningful Problem Tasks

Overview of Theme. Recent recommendations (e.g., National Council of Teachers of Mathematics, 1989, 1991; National Research Council, 1990) highlight the need for relevant, problem-driven instruction. A central thesis of these recommendations is that *all* students should become confident "doers" of mathematics and, consequently, be capable and resourceful problem solvers. This requires that *all* students have the opportunity to explore many different types of mathematical problems and that they be both expected and encouraged to use a variety of strategies in solving them (National Council of Teachers of Mathematics, 1989).

Although problem solving traditionally has been a difficult area for many students with learning disabilities (Montague & Bos, 1986; Wansart, 1990), Bulgren and Montague (1989) reported that these students can succeed beyond current expectations *if* they are exposed to developmentally appropriate, meaningful problem tasks that are complemented by appropriate instructional modifications. Moreover, children experiencing difficulties with formal computation or basic fact recall should not be prohibited from engaging in more challenging problem-solving tasks (Cawley & Miller, 1989; Ginsburg, chapter 2 of this book). In fact, a substantial body of research highlights the effectiveness of using problem solving as the vehicle for learning mathematics, including basic facts and computation (e.g., Carpenter & Moser, 1984).

When problem tasks are sufficiently complex, rich, and open-ended, they can be explored at different levels of understanding. Stenmark (1991) characterized a "rich" problem in three ways: (a) The problem leads to other problems, (b) the problem raises other questions, and (c) the problem has many solution approaches. We would add a fourth criterion: that a problem makes multiple connections.

Case Illustration. One example of a complex, rich problem task, which was presented to a self-contained class of nine students classified as severely learning disabled, is the following triangle–rectangle problem:

> *Is every triangle 1/2 of a rectangle?*
> *Yes or no? Prove it.*

The following excerpt from the teacher's journal provides insight into both the nature of the problem task and the thinking and physical representations the students used to solve it.

> Three boys, working together, cut out the colored triangles [from the Triangle Worksheet (TWS; see Figure 7.2)], taped them to the triangles on the white TWS [See Figure 7.3], and formed parallelograms. Their premise was, "No— two equal triangles do not form rectangles. The shapes formed are not rectangles because they do not have 90 degree angles." (Stone, 1993, p. 54)

> A [second] group of two boys gave themselves permission to cut the . . . triangles [see Figure 7.3] on the altitude and tape the two [colored] triangles, one on either side, to the white triangle. They had a little difficulty with #3, the obtuse triangle. They cut off a piece along a line at the end. After they taped the two

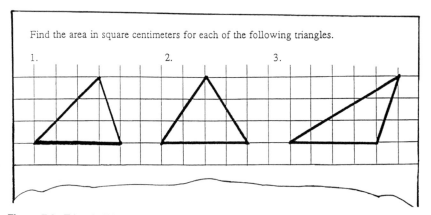

Figure 7.2. Triangle Worksheet.

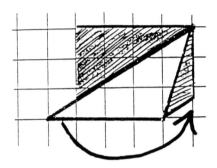

Figure 7.3. Two boys' solution to the triangle–rectangle problem.

pieces to the existing triangle they had a small piece sticking out on the left and a small hole on the right. They asked if it was O.K. to cut off the piece and move it. They ended up with a perfect rectangle with a base of 4 and a height of 3 (very ingenious!). (Stone, 1993, p. 56)

One girl was working by herself due to absenteeism. She also cut the cut-out triangles. She worked totally independently. Her first conjecture was that all triangles except #3 could form rectangles. She was very proud of herself when she finally figured out how to do #3 like the other boys did. (Stone, 1993, p. 54)

Within problem contexts such as this, students with learning disabilities are able to draw upon their diverse strengths as they solve problems using different parameters and achieve success "within their specific limitations" (Borasi, in press). This kind of exploration addresses the broader need to challenge students to think beyond common expectations.

When viewed according to the characteristics proposed by Stenmark (1991), the triangle–rectangle problem meets the criteria for a rich problem in that it (a) generated extension problems, (b) raised questions about shapes, (c) generated different solutions by redefining the parameters of the problem, and (d) set the stage for exploring further connections.

In relation to extension problems, the group of children in the first scenario just mentioned correctly reasoned that every triangle *is not* half a rectangle but *is* half a parallelogram. This raised a further problem that was pursued in a later lesson, "Is every triangle half a parallelogram?" The second group of three boys redefined the problem in their own way and, in essence, investigated an extension problem: "Can a rectangle be formed by physically changing two congruent triangles?"

The problem raised questions about the defining properties of shapes. For example, when is a parallelogram a rectangle? The problem also provided an opportunity for the teacher to follow up on the distinction between congruent shapes and shapes that have the same area.

Because of differing interpretations and reasoning, the problem gave rise to two different but valid solutions. In one case, assuming that the shape of the triangle could not be changed, the children concluded that it was not possible for every triangle to be half a rectangle. In the other case, the children made a different assumption—that the shape of the triangle could be changed as long as areas remained the same. In this situation it was possible to construct a rectangle that was twice the area of the given triangle.

In terms of connections, the triangle–rectangle problem task set the stage for the teacher to make the link between areas of triangles and areas of rectangles. The natural connection between the visualization of length–width measures of a rectangle and the corresponding base–height measures of a triangle could be highlighted in such instruction. Further, it would be possible to make connections among the areas of a triangle, a rectangle, and a parallelogram.

When students are given ongoing opportunities to engage in rich problem tasks, as in this case, the results can be quite dramatic. This success is consistent with research documenting the fact that students learn what they have an opportunity to practice. Students who have had many opportunities to solve mathematical problems become better at problem solving (e.g., Carpenter et al., 1989; National Council of Teachers of Mathematics, 1989; Silver, 1985).

Accommodating the Diverse Ways in Which Children Learn

Overview of Theme. Mathematics today is viewed as a "sense-making experience" involving numerical, logical, and spatial concepts and relationships. Because sense-making is idiosyncratic, students with learning disabilities generally need considerable time to understand problem situations and construct strategies. Further, if these students are to develop higher levels of mathematical thinking and more positive dispositions toward mathematics, they need ongoing opportunities to explore mathematical tasks in ways that match their learning strengths (Speer & Brahier, 1994). For example, learning groups might be formed on the basis of complementary learning styles. With this approach, students with different strengths can find their niches and achieve success within their specific limitations (Borasi, in press).

Case Illustration. An episode involving Dan (a pseudonym; Behrend, 1994) is a case in point. Dan, a 9-year-old, received mathematics instruction in a learning disabilities resource room. On the WISC-R, his Full Scale IQ score was average, although he had difficulty processing multiple pieces of information. At the time of the study he was on medication to control his attention deficit disorder (ADD).

Dan was the most inconsistent student with respect to mathematical performance in Behrend's (1994) study. His inconsistency was readily apparent in routine computational problems, where he attempted to apply learned rules in a nonmeaningful way. For example, when asked which of two ways (see Figure 7.4a) would be the better way to find the total, Dan selected the example on the left because it corresponded to his interpretation of the teacher's rule for adding: "The ones are first" (p. 74). Dan believed that 78 was a reasonable answer because the 4 was "where it's supposed to be . . . . Because that's where you mostly put the first number for numbers" (p. 75).

However, when Dan was faced with a nonroutine problem and was allowed the flexibility to solve it in his own way, he demonstrated surprising insight, as illustrated by his solution to the following problem.

> Nineteen children are taking a bus to the zoo. They will have to sit either two or three in a seat. The bus has 7 seats. How many children will have to sit three to a seat, and how many can sit two to a seat? (Behrend, 1994, p. 77)

Dan quickly drew seven lines to represent seats, drew a circle for each seat, and repeated the process until he had accounted for 19 circles (see Figure 7.4b). This kind of modeling and counting strategy exemplifies Dan's thinking in problem situations for which a known procedure was not readily available. Not only was he capable of solving nonroutine problems like this, but Behrend reported that he was also capable of correctly solving problems that included extraneous information.

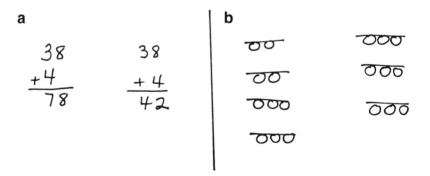

Figure 7.4. Dan's work.

In effect, when the teacher accommodated Dan's distinctive learning style, Dan was successful; when he felt compelled to use the teacher's algorithmic approach, he invariably failed. The inflexibility of a meaningless procedure appeared to inhibit his ability to recognize the reasonableness of an answer or his attempts at alternative strategies.

As this case illustrates, accommodating the diverse ways in which children learn does not always require proactive strategies on the part of the teacher. Rather, there are times when the teacher needs to step back and observe and listen to children's thinking patterns so that he or she can respond to and maximize the children's strengths.

In her study, Behrend (1994) found that students with learning disabilities constructed and utilized their own strategies to solve a wide variety of problem types. She concluded that instruction should build on children's current understandings and promote the development of increasingly more efficient problem-solving strategies, *rather than emphasizing specific rules and procedures.* In finding implications for instruction, Behrend generates a powerful message for teachers of diverse learners:

> A model of instruction which involves posing problems, allowing students time to solve the problems in their own way, listening to students' strategies, assisting only when necessary, and discussing similarities and differences between strategies provides several advantages over the other forms of instruction. Teachers are able to make assessment an integral part of instruction, students are given more control over their learning, and mathematics is seen as a process of making sense of number relationships. Instruction becomes less a matter of following directions, or imitating what has been modeled, and more a way of making connections to what is already known. (p. 109)

Encouraging Students to Discuss and Justify Their Problem-Solving Strategies and Solutions

Overview of Theme. Research has suggested that classrooms in which students "discuss, critique, explain, and when necessary, justify their interpretations and solutions" (Cobb et al., 1991, p. 6) are effective at nurturing mathematical thinking. Such inquiry-oriented approaches are consonant with Scheid's (1990) review of research in special education, which has also emphasized the importance of children's thinking about their solutions and justifying them.

Students can communicate and justify their thinking and reasoning through journal writing, partner sharing, or whole-class discussion, depending on the situation and individual student needs. After completing a problem task, teachers could invite students to share their thinking or journal entry with a partner or small group. In this way, all students have an opportunity to communicate their thinking in some way, whether or not they subsequently share their ideas with the larger group. This Think-Pair-Share approach (McTighe & Lyman, 1988) increases the kinds of personal communications that are necessary for students to internally process, organize, and retain ideas (Pimm, 1987).

Whole-class discussions in which students explain and justify their solutions to problems provide a rich forum in which students develop their understanding of mathematics. In sharing their ideas, students take ownership of their learning and negotiate meanings, rather than relying solely on the teacher's authority (Cobb et al., 1991). Lo, Wheatley, and Smith (1991) also reported positive changes in students' dispositions and self-esteem when they were expected to listen to each other and respect others' ideas.

Students with diverse learning needs gain credence with their peers by reporting to the whole class what they have learned from participating cooperatively in group work or journal writing. Reporting sessions also provide opportunities for less articulate students to learn from their peers who, in a sense, serve as role models for higher level thinking. Repeated exposure to such experiences enhances the likelihood that students with specific disabilities will begin to independently think at higher levels (Scheid, 1990).

Case Illustration. Borasi, Kort, Leonard, and Stone (1993), reporting on a class of nine children with severe learning disabilities, noted how these students frequently wrote to explain to others what they did, and then paired up for sharing. In fact, the two children who also had ADHD were often invited to share their thinking while walking down the hall so as to "get rid of some excess energy" (p. 143).

In another example from Borasi et al. (1993), students were asked to write a journal entry describing their processes for finding the number of tiles needed to tessellate the classroom floor. One student, whom we will call Todd, had a severe motor disability in writing as well as a "numerical" disability. He was helped by Borasi, a participant observer in the class, to first reconstruct and then record his solution in a journal. Prompted by the researcher's questions, Todd explained how he had solved the tiling problem. He declined her offer to scribe for him, preferring to do it himself. He described each step of his solution process aloud before writing it down. The journalizing task was completed over a 2-day period, with support and questioning from Borasi. In the end Todd produced a well-organized and understandable entry, which Borasi later transcribed on a computer for sharing with other students.

Reflecting on Todd's experience, the researcher commented that

> this one-on-one work seemed really important and productive with a student with severe writing disabilities such as [Todd]; it was our hope that this experience would show [Todd] what he could really do, and provide a model for the future; we do not expect him now to be able to do similar writing on his own yet, but perhaps he might be able to do it a second time around with less help, and gradually learn to do the same without the adult support. (Borasi et al., 1993, p. 152)

CONCLUDING COMMENTS

Consistent with recent recommendations of the National Council of Teachers of Mathematics (1989, 1991), this chapter has presented and illustrated four promising themes for mathematics instruction that have emerged from recent mathematics studies involving students with LD. These themes—(a) providing a broad and balanced mathematics curriculum; (b) engaging students in rich, meaningful problem tasks; (c) accommodating the diverse ways in which children learn; and (d) encouraging students to discuss and justify their problem-solving strategies and solutions—suggest ways for rethinking the teaching and learning of mathematics for students with learning disabilities.

Case data exemplifying these themes provide a vision of what can happen when teachers nurture mathematical thinking and provide time and opportunity for students to engage in and share their solutions to rich, meaningful problems. Students with cognitive and processing disabilities deserve—and have the potential—to be empowered mathematically. In the field of learning disabilities, relatively few studies reflect the instructional themes identified in this chapter. This chapter has highlighted four studies that illustrate successful engagement of students with LD in problem solving and higher level thinking. Consistent with the findings of these studies, we are recommending a broadened approach to curriculum and instruction that accommodates and capitalizes on diversity in thinking and learning.

Although further research is needed, the studies outlined in this chapter suggest that the mathematical abilities of students with LD can be accommodated and capitalized upon when these students have pervasive opportunities to learn in challenging, broad, and well-balanced programs. The fact that students may need appropriate compensatory techniques notwithstanding, our thesis is that programs based on the themes presented in this chapter can raise the mathematical thinking of these students to levels previously considered to be beyond their reach.

> *If you don't let your grasp extend your reach, then you'll never extend your reach.*
> —Woody Allen, 1992

8. Mathematics Instruction for Secondary Students with Learning Disabilities

ERIC D. JONES, RICH WILSON, AND SHALINI BHOJWANI

The body of research on mathematics instruction for secondary students with learning disabilities (LD) is not developed well enough to describe a specific and comprehensive set of well-researched practices, but it is sufficient for defining a set of procedures and issues as clearly associated with effective instruction and increased student achievement. In this chapter, data-based investigations of procedures that have evaluated the effectiveness of mathematics instruction with secondary students with LD will be discussed. Although this discussion will be based primarily on studies that were limited to secondary students with LD, some research on the instruction and achievement of younger students or higher achieving populations will be considered.

This discussion considers six factors that predictably confound efforts to increase the effectiveness of instruction. Each of the factors is particularly relevant in the case of instruction for secondary students with LD. These factors are (a) students' prior achievement, (b) students' perceptions of self- efficacy, (c) the content of instruction, (d) management of instruction, (e) educators' efforts to evaluate and improve instruction, and (f) educators' beliefs about the nature of effective instruction.

PRIOR ACHIEVEMENT

Although students with LD spend a substantial portion of their academic time working on mathematics (Carpenter, 1985), severe deficits in mathematics achievement are apparent and persistent. Although secondary students with LD continue to make progress in learning more complex mathematical concepts and skills, it appears that their progress is very gradual (Cawley, Fitzmaurice, Shaw, Kahn, & Bates, 1979; Cawley & Miller, 1989). McLeod and Armstrong (1982) surveyed junior high, middle school, and high school math teachers regarding mathematics achievement. The teachers reported that skill deficits in basic computation and numeration were common. Specifically, McLeod and Armstrong found that secondary students with LD had difficulty with basic operations, percentages, decimals, measurement, and the language of mathematics. Algozzine, O'Shea, Crews, and Stoddard (1987) examined the results of 10th graders who took Florida's minimum competency test of mathematics skills. Compared to their general class peers, the adolescents with LD demonstrated substantially lower levels of mastery across all subtests. Algozzine et al. reported that the students with LD consistently scored higher on items requiring the literal use of arithmetic skills than on items requiring applications of concepts. Similarly, the results of the National Assessment of Educational Progress (cited in Carpenter, Matthews, Linquist, & Silver, 1984) clearly indicated that too many students in the elementary grades failed to acquire sufficient skills in operations and applications of mathematics. These persistent skill deficits, combined with limited fluency of basic fact recall (i.e., lack of automaticity), will hinder the development of higher level mathematics skills and will compromise later achievement (Hasselbring, Goin, & Bransford, 1988).

For secondary students with disabilities, the adequacy of instruction in mathematics will be judged not merely on how quickly basic skills can be learned. Students must also acquire generalizable skills in the application of mathematical concepts and problem solving. The task of designing instructional programs that result in adequate levels of acquisition and generalization for students who have experienced seriously low levels of achievement for a large proportion of their academic careers is indeed a challenging one.

PERCEPTIONS OF SELF-EFFICACY: ATTRIBUTES FOR FAILURE OR SUCCESS

Individual differences in cognitive development certainly affect the achievement of academic skills. In earlier years, many professionals readily accepted that individual psychological differences accounted for failure to learn in school. Currently, a more parsimonious explanation is that many students fail as a result of ineffective instruction (Engelmann & Carnine,

1982; Kameenui & Simmons, 1990). Students' expectations for failure frequently develop as a result of prolonged experiences with instruction that fails to result in successful performance.

By the time students with LD become adolescents, they have typically endured many years of failure and frustration. They are fully aware of their failure to achieve functional skills in the operations and applications of mathematics. Although research supports the argument that perceptions of self-efficacy are task specific and fairly accurate, it does not reveal that learning disabilities are consistently associated with lower general self-concepts (Chapman, 1988). Chapman concluded that students who come to doubt their abilities (a) tend to blame their academic failures on those deficits, (b) generally consider their low abilities to be unchangeable, (c) generally expect to fail in the future, and (d) give up readily when confronted with difficult tasks. Unless interrupted by successful experiences, continued failure tends to confirm low expectations of achievement, which in turn sets the occasion for additional failure.

Pajares and Miller's (1994) study of self-efficacy and mathematics has implications for teachers who attempt to remediate low math achievement. They found that students' judgments of their ability to solve specific types of mathematics problems were useful predictors of their actual ability to solve those problems. Specific student estimates of self-efficacy were more accurate predictors of performance than prior experience in mathematics. Extending the results suggests three important issues. First, judgments of self-efficacy are task specific and generally accurate. Second, student judgment of self-efficacy may provide insights that will be valuable supplements to teacher assessments of performance skills. Third, negative expectations and motivational problems may be reduced by interventions to eliminate deficits in specific mathematics skills. The fact that student ratings of self-efficacy are accurate (for both successful and unsuccessful students) suggests that this is not a chicken-or-the-egg–type issue. Instead, it suggests that initially there is a fairly direct path from instruction to performance and, subsequently, to perceptions of self-efficacy. When instruction is effective (i.e., when students master targeted competencies), performance is enhanced and an accurate and positive perception of self-efficacy results. Conversely, ineffective instruction leads to poor performance and an accurate but negative estimate of self-efficacy.

For secondary students with LD, expectations to fail to learn mathematics skills can be an important obstacle. Although students' perceptions of their self-efficacy originally develop from their experiences with success and failure in instruction, those expectations later can become factors that prevent low-achieving students from attempting to learn, or persevering in trying to apply, mathematical concepts and skills.

CONTENT OF INSTRUCTION

With high-quality instruction, students will acquire skills in less time and make more adaptive generalizations than they would with lower quality instruction. Quality of instruction is dependent on two elements of curriculum design: organization of content and presentation of content. Secondary students are expected to master skills in numeration and mathematical operations and to be able to apply those skills across a broad range of problem-solving contexts. Adolescents with LD are not likely to acquire adequate competence unless the content of their instruction is carefully selected and organized. Woodward (1991) identified three empirically supported principles related to curriculum content that contribute to the quality of instruction: the nature of examples, explicitness, and parsimony.

Nature of Examples

Students learn from examples. An important part of the business of education is selecting and organizing examples to use in instruction such that students will be able to solve problems they encounter outside of instruction. Unfortunately, commercial math curricula frequently do not adequately manage the selection or organization of instructional examples. Two deficiencies that contribute to inefficient instruction and chronic error patterns in the management of instructional examples are common to commercial math curricula. First, the number of instructional examples and the organization of practice activities are frequently insufficient for students to achieve mastery (Silbert, Carnine, & Stein, 1990). As a result, although high-achieving students may quickly attain near-perfect performance, low-achieving students (including students with a variety of learning difficulties) fail to master the same math skills before teachers move on to new instructional tasks. Empirical studies of two types support the validity of this phenomenon. First, classroom observations of teacher behavior indicate that teachers tend to direct their instruction (in terms of the difficulty of the material) to students of high-average achievement (Brophy & Good, 1974). Second, classroom observation indicates that when students with LD are taught in mainstream settings with students without disabilities, they average only 60% correct (Chow, 1981), a figure considerably below the performance levels of 90% to 100% correct that most educational experts require for task mastery.

A second deficiency is an inadequate sampling of the range of examples that define a given concept. If some instances of a concept are underrepresented in instruction or simply not included in instruction, students with LD will predictably fail to learn that concept adequately. The direct connection between the range of examples and task mastery has been demonstrated in instructional areas including teaching fractions (Kelly, Gersten, & Carnine, 1990) and test taking (McLoone, Scruggs, Mastropieri, & Zucker, 1986). The adequacy of a selection of examples depends on several factors, including (a) possible variations of the concept, (b) the likelihood that irrelevant or misleading variables will be erroneously associated with the concept, (c) the complexity of the concept being taught, and (d) the variety of potential applications of the concept.

If inadequate selections of instructional examples are provided, the range of a concept is not illustrated and students form limited or erroneous conceptualizations. Silbert et al. (1990) provided an example of instruction in the analysis of fractions that predictably contributes to limited understanding of fractions: Generally, students are taught that fractions represent equal divisions of one whole, and during the elementary grades, considerable instruction focuses on teaching the concept of a fraction, with examples such as 1/2, 1/4, 1/8, and so forth. This representation of the concept of a fraction is inadequate, however, because not all fractions are less than one whole unit. Some fractions are equal to or greater than 1 (e.g., all improper fractions, such as 6/5, where the numerator is larger than the denominator). Representing the concept of a fraction as a quantity less than one whole limits student understanding of the wider range of possible fraction concepts. An inadequate conceptualization of fractions contributes to inadequate understanding of computation with fractions and, thus, severely limits problem-solving skills. For example, students who "learn" that fractions represent quantities less than one unit may experience difficulty understanding that even the quantity represented by a proper fraction (i.e., one in which the numerator is smaller than the denominator) may not be less than 1. Consider an example involving the fact that 3/5 of the students in a class of 25 are female. In this case, the quantity represented by 3/5 of the class equals 15 individuals. To ensure that this example of the concept of a fraction is understood, teachers must include examples whereby fractions represent portions of groups as well as portions of single objects.

To learn correct conceptualizations, students must be taught which attributes are relevant and which are irrelevant. If sets of instructional examples consistently contain attributes that are irrelevant to a concept, then students will predictably learn misconceptualizations that may seriously hinder achievement. It is not uncommon to find that presentations of misleading variables have inhibited mathematics achievement. One important illustration is the frequent use of key words in story problems (e.g., Carpenter et al.,

1984; Nesher, 1976; Wright, 1968). Although so-called key words frequently appear in story problems and often indicate the solution, they are not relevant per se to the solution of the problems. Unfortunately, many low-achieving students learn to depend on key words instead of attending to the more critical information presented in the problem. Consequently, they are apt to experience difficulty solving problems that do not have key words, or that use key words in ways that are irrelevant to any mathematical solution.

Such misconceptualizations will confuse students unless their exposure to potentially misleading cues is carefully managed. Initial sets of instructional examples should not contain key words. As students become proficient with those examples, problems that contain relevant key words can be included in instructional sets, juxtaposed with examples in which the key words are irrelevant to the solution (see Engelmann & Carnine, 1982; Kameenui & Simmons, 1990; Tennyson & Park, 1980).

Secondary students must learn to deal with complex notations, operations, and problem-solving strategies. Complexity is sometimes related to the level of abstraction the student must deal with; it may also be related to understanding the relationships between associated concepts. Thus, instuctional examples should provide for the systematic progression from concrete to more abstract representations (Mercer & Miller, 1992) and from simpler to more involved relationships among concepts and rules. Students with LD have difficulty with complex tasks because they do not receive sufficient opportunities to work with complex instructional examples. Although instruction on simple forms of concepts or strategies does not facilitate generalization to more complex forms without additional instruction, Rivera and Smith (1988) demonstrated that carefully managed instruction for solving more complicated forms of division problems facilitated students' independent solutions of simpler division problems.

Explicitness

Explicitness of curriculum design refers to the unambiguous presentation of important concepts and skills and the relationships among them (Woodward, 1991). The explicitness of a curriculum is affected by the quality of teacher decisions made and actions taken at the following five stages of the instructional design process:

- Determining the concepts and skills that must be learned;
- Identifying the important relationships among concepts and skills;
- Organizing facts, concepts, and skills into logical hierarchies;

- Developing sets of instructional examples that unambiguously illustrate the range of concepts and skills that must be mastered; and
- Presenting the instructional examples to the student.

Ambiguities that enter the design process will result in predictable misunderstandings. Although research in this area is limited, Jitendra, Kameenui, and Carnine (1994) demonstrated that a highly explicit math curriculum produced greater student achievement than a less explicit curriculum.

The premise that curriculum quality is related to the degree to which concepts and skills are explicitly taught is being debated in the current movement to reform mathematics education. Instead of the argument for unambiguous organization and presentation, an increasingly popular position is that students should participate in instructional activities that allow them to construct both knowledge of concepts and skills and understandings of interrelated hierarchies (e.g., Poplin, 1988a, 1988b). Others contend that although students construct their own knowledge, teachers can contribute to the efficiency of instruction by carefully planning and structuring the learning experience (e.g., Harris & Graham, 1994; Mercer, Jordan, & Miller, 1994; Pressley, Harris, & Marks, 1992). Engelmann (1993) acknowledged that it is logically impossible to have universal hierarchies of instructional skills, but he also contended that curricula must be organized around explicit instructional priorities. According to Engelmann, if hierarchical sequences are not developed around explicit instructional priorities, it is unlikely that students with learning difficulties will progress efficiently.

We encourage practitioners to follow this debate in the professional literature, because there is likely to be a subsequent effect on classroom practice. However, for students with LD, we believe that this dictum accurately summarizes the empirical literature: More explicit instruction results in more predictable, more generalizable, and more functional achievement. If we do not explicitly teach important knowledge and skills, these objectives will not be adequately learned.

The need for explicitly identifying and teaching important mathematical concepts, skills, and relationships is apparent in the persistent failure of students with LD to deal adequately with common fractions, decimal fractions, percentages, ratios, and proportions. A clear example of the link between explicit instruction and expected student performance can be seen in states that use proficiency tests as criteria for promotion or graduation. In the state of Ohio for example, students in the ninth grade take basic academic-proficiency examinations. They must pass these exams in order to graduate from high school. The Ohio Department of Education provides funds for school districts to deliver remedial programs to students who fail the proficiency examination. The remedial programs provide explicit instruction on the types of problems and relationships presented on the test. Although one might question the wisdom or utility of proficiency testing, the telling argument here is that when educators were forced to confront the practical problem of designing the most efficient and effective means for students to achieve criterion performance (i.e., pass a test), an explicit approach was selected.

Parsimony

Effective curricula provide for an economical, or parsimonious, use of time and resources. Woodward (1991) contended that emphasis should be given to mastery of concepts, relationships, and skills that are essential for the subsequent acquisition and functional generalization of math skills. Curricula should be organized so that instruction of specific skills and concepts is tightly interwoven around critical concepts. Woodward's test for the parsimony of an instructional program is whether or not what is learned at one time will be used later. In general, practitioners are well aware of the need for functional instruction, because they must frequently answer questions such as, "Which is more important to the students, a unit on metric conversion or one on checking account management?" or "How often in life will students with LD be required to divide fractions?" However, it is not enough to identify information and skills that take the highest priority. The curriculum must be organized so that the greatest amounts of high-priority information and skills can be mastered as efficiently as possible.

In summary, the content of instruction and its organization play critical roles in determining its quality and outcomes. Unfortunately, commercial math curricula frequently stop well short of providing adequate opportunities to learn to solve mathematics problems that involve the contexts of work and everyday situations. Successful student performance on algorithms and abstracted word problems does not always result in competent real-life mathematical problem solving. In response to that problem, Hasselbring and colleagues (Bottge & Hasselbring, 1993; Bransford, Sherwood, Hasselbring, Kinzer, & Williams, 1990; Cognition and Technology Group at Vanderbilt University, 1991; Hasselbring et al., 1991) have been conducting innovative research on teaching adolescents with LD to solve contextualized mathematical problems. They use videodisks and direct instruction techniques to present complex mathematical problems that are embedded in portrayals of real-life situations. The results of their studies are encouraging and clearly demonstrate that adolescents with LD can be taught to solve more complex problems than their teachers generally expect from them. On the other hand, those studies also demonstrate that unless students are guided in solving complex math problems and are also given sufficient opportunities to independently attempt to solve such problems, they usually will not learn adequate generalized problem-solving skills.

MANAGEMENT OF INSTRUCTION

Archer and Isaacson (1989) listed three variables that can be measured to evaluate quality of instruction at any grade level: time on task, level of success, and content coverage. According to their perspective, good teachers manage instruction so that students (a) spend the major portion of instructional time actively engaged in learning, (b) work with high levels of success, and (c) proceed through the curriculum while acquiring increasingly more complex skills and important generalizations. Thus, good teaching is indicated by students' responses to instruction. Obtaining high levels of achievement requires effective management of instruction.

Zigmond (1990) reported data from observational studies indicating that too many teachers of secondary students with LD may not be managing instruction effectively. She reported that during their class periods, resource room teachers spent slightly less than 40% of class time in instructional interactions. Teachers spent 28% of their time "telling students what to do, but not teaching them how to do it, and another 23% of the time not interacting with students at all" (p. 6). Although students were observed to be on task for about three quarters of the class period, they were often completing worksheets. Worksheets may be appropriate for practice, but they are not useful for introducing new information and skills. The lecture format of instruction that often occurs in general secondary mathematics classes often is not effective, either. Zigmond observed that many large classes are not managed well—students who have difficulty understanding what the teacher is talking about tend to be off task, and misunderstandings often go undetected.

Alternatives to worksheet instruction and didactic lectures have been investigated in empirical studies. The approaches to managing instruction that we will discuss are *direct instruction, interactive instruction, peer-mediated instruction,* and *strategy instruction.* The effectiveness of the first three of these approaches has been documented across a variety of curriculum areas with secondary students in general, remedial, and special education programs.

Direct Instruction

There are a variety of interpretations of the term *direct instruction*. In some discussions it refers to a model for both curriculum organization and presentation procedures (e.g., Becker & Carnine, 1981; Engelmann & Carnine, 1982; Gersten, 1985). Other descriptions of direct instruction refer to a set of procedures for actively involving students in academic learning (Christenson, Ysseldyke, & Thurlow, 1989; Rosenshine, 1976; Rosenshine & Stevens, 1986). In both discussions, instruction is teacher-led and characterized by (a) explicit performance expectations, (b) systematic prompting, (c) structured practice, (d) monitoring of achievement, and (e) reinforcement and corrective feedback.

Archer and Isaacson (1989) provided a structure for teacher-led instruction that is divided into three phases: the lesson opening, the body of instruction, and the closing of the lesson (see Table 8.1). During each phase the teacher works to maintain high levels of active student involvement, successful acquisition, and progress through the curriculum.

The Opening of the Lesson. The teacher first gains the attention of the students. A brief statement, such as, "Look up here. We are going to begin," is generally adequate. The teacher reminds the students of what was accomplished in the previous lesson and sets the goal for the current session. For example, "Yesterday we learned how to calculate the area of squares and rectangles. Today we are going to learn how these calculations are used in the home to lay carpet, paint walls, and tile floors." Lengthy reviews or previews

---

Table 8.1
Phases of Structured Academic Presentations

Opening
  Gain the students' attention.
  Review pertinent achievements from previous instruction.
  State the goal of the lesson.

Body
  Model performance of the skill.
  Prompt the students to perform the skill along with you.
  Check the students' acquisition as they perform the skill independently.

Close
  Review the accomplishments of the lesson.
  Preview the goals for the next lessons.
  Assign independent work.

---

of upcoming lessons are unnecessary. If this part of the lesson is not brief, students are likely to start attending to other things.

The Body of the Lesson. It may be necessary to cue the students' attention again with a gesture or brief comment before beginning, but immediately after opening the lesson the teacher should begin instruction with the first example. With the first phase of instruction the teacher models the task. For example, if the students are learning a problem-solving strategy, the teacher first models the strategy by saying, "If I wanted to determine how many 1' by 1' self-stick floor tiles it would take to cover an 8' by 10' floor, I could apply the surface area formula we learned yesterday. I would multiply 8' times 10' and find that it would take 80 square feet. If I know that the room has 80 square feet, how many 1' by 1' floor tiles do I need?" The demonstration step is critical, especially if students are learning a new concept or skill. It should be brief and should explicitly indicate the critical aspects of applying the strategy. After the demonstration step, the students are led through a few examples. In this "lead step," the students and teacher actively respond together to solve the problem, introduced by stating, "Let's try a problem together." Students may respond orally, in writing, or by gesture. Archer and Isaacson (1989) recommended that teachers use various forms of student responses.

Several prompted trials are usually necessary before students can be expected to respond independently. The transition from prompted trials to independent trials can take place with few or no errors if teachers systematically delay the delivery of a prompt (Cybriwsky & Schuster, 1990; Koscinski & Gast, 1993; Koscinski & Hoy, 1993; Stevens & Schuster, 1988).

After students respond correctly on several independent test trials, the teacher should have them complete practice examples. If the students are practicing newly acquired skills, they should be supervised for at least the first few practice examples. If they do not need assistance completing their practice exercises, the teacher may close the lesson.

The literature on direct instruction of students with LD identifies five recommendations in the delivery of instruction that contribute to the effectiveness of that instruction: (a) Obtain frequent active responses from all students, (b) maintain a lively pace of instruction, (c) monitor individual students' attention and accuracy, (d) provide feedback and positive reinforcement for correct responding, and (e) correct errors as they occur (Archer & Isaacson, 1989; Becker & Carnine, 1981; Christenson et al., 1989; Mercer & Miller, 1992; Reith & Evertson, 1988). Each recommendation makes an important contribution, but achievement will be highest when all five are part of the delivery of instruction (Becker & Carnine, 1981).

The Close of the Lesson. Typically, the teacher closes a lesson with three brief steps. First, he or she reviews what was learned during the current lesson, where there may have been difficulty, and where performance may have been particularly good. The review may also include a brief statement of how learning in the current session extended what was already known. Second, the teacher provides a brief preview of the instructional objectives for the next session. Third, she or he assigns independent work. Independent seatwork and homework provide important opportunities for students to apply knowledge and practice skills that they have already learned, thus increasing fluency and retention. Independent practice should, therefore, be carefully selected so that students can actually complete it successfully without assistance from a teacher or parent.

Practice activities are essential components of mathematics instuctional programs. Students with LD will generally need more practice and practice that is better designed than students without LD, if they are to achieve adequate levels of fluency and retention. Worksheets are commonly used to provide practice, but the ones that publishers supply are frequently inadequate. Table 8.2 provides a list of principles for designing and evaluating practice activities for students with LD. For a more detailed discussion of those guidelines, see Carnine (1989).

Table 8.2
Principles for Designing Practice Activities

---

1. Avoid memory overload by assigning manageable amounts of practice work as skills are learned.
2. Build retention by
   • providing review within a day or two of the initial learning of difficult skills, and by
   • providing supervised practice to prevent students from practicing misconceptions and "misrules."
3. Reduce interference between concepts or applications of rules and strategies by separating practice opportunities until the discriminations between them are learned.
4. Make new learning meaningful by
   • relating practice of subskills to the performance of the whole task, and by
   • relating what the student has learned about mathematical relationships to what the student will learn about mathematical relationships.
5. Reduce processing demands by
   • preteaching component skills of algorithms and strategies, and by
   • teaching easier knowledge and skills before teaching difficult knowledge and skills.
6. Require fluent responses.
7. Ensure that skills to be practiced can be completed independently with high levels of success.

Note. Adapted from Carnine (1989).

Model–lead–test presentation procedures have many positive qualities: They are quite easy to learn; they are, with slight variations, generalizable to teaching motor skills, concept discriminations, rule relationships, and strategy tasks (see Archer & Isaacson, 1989; Kameenui & Simmons, 1990; Silbert et al., 1990). By themselves, the presentation procedures have been demonstrated to provide for efficient instruction across age levels, ability levels, and curriculum domains. If students imitate the initial prompts, there is little chance that they will guess and make errors. Because prompts are faded, the chances that students will develop dependence on instructional prompts is reduced. Thus, the students learn to expect to be successful and are less apt to resist instruction than if they are confronted with loosely arranged instructional procedures where the chances that they will fail are high. Finally, there is empirical evidence in the professional literature that direct instruction procedures have been effectively used to teach math skills to older students with LD (e.g., Perkins & Cullinan, 1985; Rivera & Smith, 1988).

Interactive Teacher Presentations and Student Responses

Outside school, the demand to demonstrate mathematics knowledge and skills emerges in a potentially wide range of representations. Appropriate responses also vary. Cawley, Fitzmaurice, Shaw, Kahn, and Bates (1978) described a model for programming mathematics instruction for secondary students with LD that takes into account both the students' skills and the various possible representations of mathematics problems. Their instructional program is interactive. Instructional examples are selected according to (a) the individual student's mastery of specific skills, (b) the mode in which the math problem will be represented, and (c) the mode in which the student will respond to the problem. Problems can be presented in four modes: active constructions or manipulation of objects, fixed visual displays, oral statements, and written or symbolic representations. The student's response mode may also occur across four options: active constructions or manipulation of objects, identification of a visual display from a set of alternative displays, an oral statement, and a written or symbolic representation.

Table 8.3 illustrates Cawley et al.'s (1978) options for presenting and responding to problems. The modes for the teacher's presentation and the student's responses can be mixed; for example, the teacher may present a problem orally, and the student may respond by selecting an appropriate response from alternatives. The 16 different combinations of modes of presentation and response allow instructional programs to be tailored to both the individual needs of students and the realistic representations of mathematical problem-solving situations. Cawley et al. argued that students who have the opportunity to work with varied presentations and response options are more likely to learn generalizable skills than students who spend the greatest proportion of their time making written responses to problems presented in workbook pages.

Peer-Mediated Instruction

Over the past 20 years, educators have taken a growing interest in using students to help with each other's academic achievement. Current initiatives in both general and special education include two major forms of peer-mediated instruction—peer tutoring and cooperative learning. Both enjoy broad support in the empirical literature. Peer tutoring can take various forms. Two classmates may take turns helping each other in one-on-one practice of skills that have been presented earlier. It could also be an arrangement whereby a higher achieving student helps, or monitors the performance of, a lower achieving student (e.g., Maheady, Sacca, & Harper, 1987). Cooperative learning also involves peers assisting each other, but, instead of pairs of students, cooperative learning groups usually comprise three or more students of differing ability levels. Slavin (1983) defined cooperative learning as instructional arrangements in which students spend much of their class time working in small, heterogeneous groups on tasks they are expected to learn and help each other learn. The aspects that contribute to the effectiveness of both approaches will be discussed in this section.

Slavin's (1983) analysis of 46 studies of cooperative learning resulted in three important conclusions regarding which variables contribute to the success of cooperative learning. First, there was no evidence that group work itself facilitated individual students' achievement. Second, cooperative incentive structures (where two or more individuals depend on each other for a reward that they will share if they are successful) themselves did not have significant effects on the achievement of individual students; group rewards for group products did not improve student achievement. Third, group incentive structures were associated with higher achievement if the performance of individual students was accounted for and was reflected in the group rewards. The simplest method for managing individual accountability is to average individual scores to determine the reward for group members. A second method is to determine group rewards based on whether, or by how much, individual members exceeded their individual criterion. A third method is to assign each student in a group a unique task. Slavin warned, however, that although task specialization provides for individual accountability, it must be

Table 8.3
Options for Presenting and Responding to Math Problems

Presentation options	Response options
Construct a problem with actual objects or manipulatives	• Construct a response by manipulating objects • Choose from an array of possible responses (i.e. multiple choice) • Make an oral response • Make a written or symbolic representation
Present a problem in a fixed visual display	• Construct a response by manipulating objects • Choose from an array of possible responses (i.e., multiple choice) • Make an oral response • Make a written or symbolic representation
Orally state the problem	• Construct a response by manipulating objects • Choose from an array of possible responses (i.e., multiple choice) • Make an oral response • Make a written or symbolic representation
Present the problem in written or symbolic form	• Construct a response by manipulating objects • Choose from an array of possible responses (i.e multiple choice) • Make an oral response • Make a written or symbolic representation

Note. Adapted from Cawley et al. (1978).

accompanied by group rewards. Furthermore, task specialization is not appropriate in situations where the instructional goal is for all members of the group to acquire the same knowledge and skills.

In summary, individual accountability with group rewards contributes to higher levels of achievement than individualized incentive structures, but group work without individual accountability or group rewards does not contribute to higher achievement than might be obtained with individualized task and incentive structures. Slavin (1983) cautioned that peer norms and sanctions probably apply only to individual behaviors that are seen by group members as being important to the success of the group. For example, he suggested that when rewards are group oriented and there is no individual accountability, group norms may apply to the behaviors of only those group members who are considered by the group as most apt to contribute to the quality of the group product.

Several studies of cooperative learning have demonstrated its effectiveness for teaching mathematics skills (e.g., Madden & Slavin, 1983; Slavin, 1984; Slavin & Karweit, 1981; Slavin, Leavey, & Madden, 1984; Slavin, Madden, & Leavey, 1984). Several of those studies (Madden & Slavin, 1983; Slavin, 1984; Slavin & Karweit, 1981) also consistently revealed that low-achieving students in cooperative learning programs enjoyed greater social acceptance by their higher achieving peers and reported higher levels of self-esteem than did low achieving students in traditional instructional programs. In one particularly relevant study, Slavin and Karweit (1984) compared the effectiveness of cooperative learning and direct instruction for teaching mathematical operations from real-life problems to low-achieving students in ninth grade. They observed higher levels of achievement with cooperative learning than with direct instruction alone. Combining direct instruction with cooperative learning procedures did not produce higher levels of achievement than cooperative learning alone. A caveat, however, is warranted, because a reasonable explanation for the relative superiority of cooperative incentives is that the difficulties the ninth graders in the study experienced were the result of their having forgotten or incompletely learned basic math skills from earlier instruction. In such cases, incentive structures probably play a more important role than the model–lead–test prompting technique. When teaching new concepts or skills, teachers would be well advised to combine direct instruction and cooperative learning.

Strategy Instruction

Students with LD must take an active role in managing their instruction. They must be able to solve problems independently, because teachers and peers are not always available or able to help them. Not only must students master the information and skills taught in their classes, but they must also successfully apply that knowledge and those skills to solve varied and often complex mathematical problems that they encounter outside instruction. Competent students independently select, apply, and monitor strategic procedures to solve complex and novel problems. A common attribute of students with LD, however, is that they do not employ effective learning strategies, unless they are explicitly instructed, and, thus, they are less effective at generalizing skills and knowledge outside the instructional setting (Deshler & Schumaker, 1986; Lenz & Deshler, 1990). If, however, students with learning difficulties are taught to use appropriate learning strategies and are reinforced for using them, they can perform effectively (Deshler & Schumaker, 1986; Ellis, Lenz, & Sabornie, 1987a, 1987b; Lenz & Deshler,

1990; Montague, Bos, & Doucette, 1991; Pressley, Symons, Snyder, & Cariglia-Bull, 1989).

Deshler and associates (e.g., Deshler & Schumaker, 1986; Lenz & Deshler, 1990; Ellis et al., 1987a, 1987b) have articulated a model for strategy instruction specifically for secondary students with LD. According to their model, effective strategy instruction follows a process that is consistent with the development of curriculum and instruction that we described in the previous sections of this chapter. Strategies are selected with reference to the curriculum demands. Teachers manage the instruction of strategies by overtly modeling strategies and then leading students through their applications. Students verbalize their applications of strategies and monitor their own progress. The teacher also provides the students with many opportunities to determine which strategies are appropriate, to use the strategies, and to be rewarded for successful applications.

In the case of mathematics, students confront many quantitative and conceptual relationships, algorithms, and opportunities to apply mathematical knowledge to solve problems. Thus, training for generalization and strategic problem solving can become a ubiquitous part of mathematics curricula. By learning to be active and successful participants in their achievement, students learn to perceive themselves as competent problem solvers. They are more apt to attempt to apply knowledge in novel ways and to persevere to solve difficult problems than if they see themselves as ineffective, likely to fail, and dependent on others to solve novel and difficult problems.

Empirical research on strategy instruction has not been comprehensive in all areas of instruction, but educators should be optimistic and make reasonable attempts to implement strategy instruction. Many studies of strategy instruction have been conducted with secondary students across a variety of academic domains. Evaluations of strategy instruction in mathematics have demonstrated its effectiveness (Carnine, 1980; Case, Harris, & Graham, 1992; Darch, Carnine, & Gersten, 1984; Hutchinson, 1993; Montague, 1992; Montague & Bos, 1986). Because those evaluations have involved only a few areas of math skills, researchers should continue to study the conditions that influence the efficacy of strategy instruction.

EVALUATION OF INSTRUCTION

It is essential that instructional interventions be evaluated frequently. The academic difficulties of secondary students with LD are diverse and complex. Current research on mathematics instruction for students with learning difficulties is not sufficiently developed to provide teachers with precise prescriptions for improving instruction. Therefore, the best educators' best efforts will frequently be based on reasonable extrapolations. Unless instructional assessments are conducted frequently and with reference to the students' performance on specific tasks, it will not be possible to use the information to make rational decisions for improving instruction. To an increasing extent, educators have come to the conclusion that traditional standardized achievement testing does not provide adequate information for solving instructional problems, and that a greater emphasis should be placed on data from functional or curriulum-based measurements (Reschly, 1992).

Assessments of instruction should provide data on individual students' progress in acquiring, generalizing, and maintaining knowledge and skills set forth in the curriculum. Curriculum-based assessment (CBA) is an approach to evaluating the effectiveness of instruction that has gained substantial attention for its value in the development of effective instructional programs. CBA can be characterized as the practice of taking frequent measure of a student's observable performance as he or she proceeds through the curriculum. As data are gathered, the measures of student performance are organized, usually graphed, and examined to make judgments of whether the student's level and rate of achievement are adequate. The teacher's reflections on the curriculum-based measures of performance and qualitative aspects of the student's performance may suggest a variety of rational explanations and potentially useful instructional interventions. If the student is making adequate progress, then it is not necessary to modify the program. On the other hand, if progress is inadequate, appropriate interventions might include devoting more time to instruction or practice, engaging the student in higher rates of active responding, slicing back to an easier level of a task, shifting instruction to tasks that are more explicitly or more parsimonously related to the instructional objective, or changing the incentives for achievement. If curriculum-based measures appear to indicate that the student is having little or no difficulty meeting instructional criteria, the teacher should consider skipping ahead to more difficult tasks.

The value of CBA as a technique for improving quality of instruction can be attributed to the effects it appears to have on teachers' instructional behavior. First, the collection of valid curriculum-based measures requires that teachers specify their instructional objectives. Efforts to identify critical instructional objectives may also lead teachers to consider how those objectives should be sequenced for instruction. Such considerations can contribute to improvement in the quality of instruction; however, Fuchs and Deno (1991) argued that measurement of subskill mastery is not necessary and that measures of more general curriculum-based measures can provide educators with reliable, valid, and efficient procedures. Second, preparations for implementing CBA are frequently accompanied by specifications of expectations for instruction, such as how many or at what rate objectives will be learned. Fuchs, Fuchs, and Deno (1985) observed that teachers who set clear but ambitious goals for their students tended to obtain higher levels of achievement from their students than teachers who set more modest goals. Third, well-planned efforts to systematically reflect on frequent observations of academic performance are likely to result in more rational assessments of skill achievement than approaches to instruction that do not rely on frequent assessments of skill performance (Fuchs, Fuchs, Hamlett, & Stecker, 1990).

In summary, CBA provides for frequent assessments of student achievement that are directly related to instructional programs. Its use appears to have the effect of more rationally relating instructional decisions to instructional objectives and student difficulties, thus contributing to increased student achievement.

Educators' Beliefs About Effective Instruction

There is no disagreement that all students, including those with LD, should be offered the most effective instruction possible. There are, however, diverse opinions among educators about the *nature* of effective instruction. It would be gratifying if empirical research played a bigger role in directing educational practice, but, frequently, beliefs and convictions play more influential roles. Sometimes educators' beliefs have positive influences on the development of instruction. For example, if a teacher believes that the source of a student's failure lies in his or her instructional experience, then the teacher may revise the curriculum, allocate more time for instruction, slice back to a less complex level of the task, or systematically and frequently collect achievement data so that the effects of instructional interventions can be monitored.

Proponents of current efforts to reform mathematics education believe that if the quality of instruction is to be improved, then many educators will have to dramatically change their perspectives on how mathematics should be taught (e.g., National Council of Teachers of Mathematics, 1989). The National Council of Teachers of Mathematics set forth the following goals for all students: (a) to learn to value mathematics, (b) to become confident in their abilities to do mathematics, (c) to become mathematical problem solvers, (d) to learn to communicate mathematically, and (e) to learn to reason mathematically. We believe that these are worthy goals and that in order to reach them, educators must examine their beliefs; however, beliefs that form the basis of the constructivist approach to mathematics education will be insufficient for guiding the development of mathematics curricula, and may be interpreted by educators as legitimizing inadequate instructional practices.

Constructivism is an ideology that is becoming increasingly popular in the current mathematics education reform movement. Its core belief is that knowledge is not transmitted directly from the teacher to the student. Instead, the learner constructs knowledge through active engagement in the process of assimilating information and adjusting existing understandings to accommodate new ways of knowing (e.g., Cobb, 1994a, 1994b; Driver, Asoko, Leach, Mortimer, & Scott, 1994; Gadanidis, 1994; Harris & Graham, 1994; Poplin, 1988a; Pressley et al., 1992). Accordingly, the world that students come to know does not exist as an independent entity outside their minds. Their knowledge of the world is actively changing according to what they already have learned and what they are currently learning.

Descriptions of the constructivist approach depict situations in which (a) teachers possess a considerable knowledge of their subject matter; (b) teachers are able to draw on that knowledge to facilitate student learning under conditions that require a great deal of extemporaneous decision making; (c) students are highly motivated to tackle difficult and challenging tasks in order to develop greater understanding; and (d) it becomes apparent over the course of the lesson that students are acquiring, developing, or constructing more sophisticated understandings than they had before the lesson began. Bereiter (1994) commented that it would be truly exceptional if all or most of these conditions could be met. He identified three substantial obstacles to meeting the conditions: (1) inadequate teacher education, (2) trial-and-error learning, and (3) difficulty with observing progress.

Teacher education is a major obstacle, but, compared to the others, it may be the easiest to overcome. Under relatively unstructured conditions in which students play very active roles in the direction of instruction, it is difficult for even the most knowledgeable teacher to facilitate learning adequately and consistently. The second obstacle is that in constructivist approaches, students often persevere through trial-and-error learning. Under such learning conditions, students with LD are apt to make many more errors than their more capable peers. If they expect to fail, they are prone to give up or to withdraw from instruction.

The last condition (i.e., effecting and acknowledging progress) will be difficult to meet, unless very simple concepts and skills are being taught. Over the course of an instructional session, it must be apparent *to both the student and the teacher* that progress is being made. Two factors limit the probability that progress will be consistently detected and rewarded: difficulty specifying objectives, and complexity of instructional conditions. The two basic tenets of constructivism—that knowledge cannot be directly transmitted from the teacher to the student, and that students actively construct their own knowledge—suggest that precise, meaningful, and measurable learning objectives may not be established. The CBA approach cited earlier suggests that as precision in measurement of student performance increases, so does student achievement.

In summary, the constructivist perspective, though intuitively appealing, is currently unsupported by empirical research and is logically inadequate for the task of teaching adolescents with LD. It is a set of beliefs that may not be able to be implemented to the satisfaction of those who promote it (see Reid, Kurkjian, & Carruthers, 1994). Unfortunately, it is also a set of beliefs that may fail to encourage the hard work of improving the quality of instruction for students with LD, and may actually be interpreted as justification for haphazard instruction. The premise that secondary students with LD will construct their own knowledge about important mathematical concepts, skills, and relationships, or that in the absence of specific instruction or prompting they will learn how or when to apply what they have learned, is indefensible, illogical, and unsupported by empirical investigations. If instructional objectives are ambiguously organized, the objectives will not be adequately addressed, and students with LD will not achieve them.

CONCLUSIONS

Secondary students with LD spend the bulk of their instructional time on very simple math skills. As a result of frequent failure, and of prolonged instruction on such simple skills, it is generally difficult to motivate them to attempt complex tasks or to persist in independent work. By the time that students graduate or drop out of school, they will have made only the most rudimentary achievements. Few will have acquired the levels of application and problem-solving skills necessary to function independently. To a great extent, improvement in mathematics education for secondary students with LD will depend on their receiving better mathematics education while they are in the elementary grades.

We strongly believe that efforts to improve mathematics education must be based on empirical research. There is already a sufficient body of research to serve as the basis for making substantial improvements in educational practice, but much more is needed. Empirical documentation of effective practice will not be sufficient to effect improvements. Compelling research on effective instructional practices is frequently ignored (Carnine, 1992). Instead, appealing but unvalidated trends, such as constructivism and discovery learning, have caught educators' attention. Those ideologies tend to be vague and allow support for haphazard and poorly designed instruction. They are logically antithetical to the existing empirical evidence on best practices for students with LD. Educational practices that are derived from ideologies must be critically evaluated—and not merely for their fit with the political sensibilities of any particular ideology, but for their effect on the achievement of children and youth.

We are also convinced that teachers must have at hand effective instructional procedures, materials, and other resources. At the present time they must do much of the work of improving mathematics education themselves. Unfortunately, few teachers have sufficient time or training to design and comprehensively evaluate math curricula (Cobb, 1988; Swing, Stoiber, & Peterson, 1988). It is the business of commercial publishers to design instructional procedures and curricula. Instead of attempting to keep up with shifts in ideologies, publishers should attempt to produce empirically defensible tools for teachers.

9. Cognitive Strategy Instruction in Mathematics for Students with Learning Disabilities

MARJORIE MONTAGUE

Mathematics learning has been reconceptualized as a constructivist process through which children construct mathematical knowledge by linking new learning to previously acquired concepts (Hiebert & Carpenter, 1992). The premise of this perspective is that individuals set goals that lead to the construction of new knowledge, which, in turn, leads to new goals and new knowledge, thus producing a spiral effect in learning (Saxe, 1991). In this view, not only cognitive development, but also social interactions, affective development, and the context for learning, are regarded as influential factors in mathematical learning.

The National Council of Teachers of Mathematics (NCTM) embraced this "new view" of teaching and learning mathematics when it established goals for instruction that focused on conceptual development, communication skills, and problem solving (National Council of Teachers of Mathematics, 1989). As a result of the NCTM's recommendations and the new view of learning, mathematics classrooms increasingly are becoming laboratories for learning, where children actively participate in solving problems. Despite these recent shifts in theoretical paradigms and instructional methodology, students with learning disabilities continue to be at risk for failure in mathematics (Parmar & Cawley, chapter 11 of this book).

177

An instructional approach that seems promising for these students is cognitive strategy instruction, the focus of this chapter. Understanding how individuals process information and acquire and apply knowledge is important for understanding cognitive strategy instruction as an approach for teaching mathematics. In this chapter, the theoretical base for strategy instruction and the characteristics of students who have difficulties in mathematics are discussed. (For a more extensive review of mathematical disabilities, see Rivera's and Ginsburg's chapters in this book.) Then, research on strategy instruction is briefly reviewed. Finally, a practical illustration of strategy instruction is presented. In that example, a cognitive approach is used to assess and teach mathematical problem solving to middle school students with learning disabilities.

A THEORETICAL BASE

Theories of instruction draw from cognitive psychology by focusing on models of learning, typical development, and individual differences in children. Developmental psychology and information-processing theory, and, more recently, theories of affective development, provide the foundation for understanding how children think and learn (e.g., Anderson, 1990; Biggs & Collis, 1982; Case, 1985; Case, Hayward, Lewis, & Hurst, 1988; Mandler, 1989; Piaget, 1952; Sternberg, 1985; Vygotsky, 1986).

Developmental-stage theory posits that learning across and within domains occurs in a series of incremental, overlapping, and integrated stages. Case (1985), a neo-Piagetian, exemplified this theoretical perspective by combining developmental and information-processing theories and also considering the interaction of cognitive and emotional factors (Case et al., 1988). The postulates governing his theoretical model have to do with the nature of children's basic intellectual structures, the stages through which the structures develop, and the process of stage transition.

Basic to Case's (1985) theory is the notion that from the age of 1 month, children are capable of some degree of voluntary control over their own cognitive and emotional experience. The cognitive structures, or schemes, that underlie this executive control have to do with representing a current state (problem representation), representing a state that has a higher affective value objectives), and representing a sequence of operations that take individuals from one state to another (strategies). Case identified four major age-related stages of executive development (sensorimotor, relational, dimensional, and vectorial operations), each having a universal sequence of three substages that become increasingly complex during maturation, with respect to the number and organization of elements they represent. Transitions across stages occur systematically for a specific domain via a process referred to as hierarchical integration of executive structures that were constructed during a previous stage but changed form and function in the higher stage. Four general processes (problem solving, exploration, imitation, and mutual regulation) regulate the constructive, or assembly, process and involve various cognitive subprocesses.

Case (1985) noted that instruction (a form of mutual regulation) plays an increasingly significant role at each stage of development. Furthermore, an individual's capability is heightened or diminished by his or her amount of short-term storage space, which, in normal development, increases developmentally in operational efficiency. Case accounted for individual differences in his theory by considering varying rates of general executive development (i.e., differences in rate of growth of short-term storage space or in efficiency of general regulatory processes). Additionally, individuals' performance on tasks in specific domains varies considerably due to differences in the efficiency of domain-specific operations or regulatory subprocesses. Viewed within this developmental framework, mathematical learning is relatively simple and concrete at the beginning stages of intellectual development and becomes progressively more complex and abstract with increasing maturity. This taxonomy extends earlier conceptualizations of cognitive development by positing consistent and repetitive intellectual structures within hierarchically arranged functioning stages.

According to this interpretation of intellectual development, mathematics learning can be traced to movement within and across different stages as learners acquire and integrate new declarative and procedural knowledge pertaining to mathematics. At the elaborated coordination level, the highest intellectual structure within each major stage, a transition from one stage to another occurs. This transition seems to be related to the learner's short-term storage space and integration of executive processes, which help an individual organize and control interactions with the environment.

Sternberg (1985) provided additional insight into the intricacies of the cognitive system. According to his theory of information processing, knowledge acquisition and performance components interact with other intrinsic mechanisms known as *metacomponents*. These higher order executive processes have a strong impact on learning ability and are instrumental in planning, monitoring, and evaluating academic tasks. Mathematics as a problem-solving activity involves the interaction of the following metacomponential functions: definition of the task, selection of lower order processes to accomplish the task, formation of a strategy by combining several lower order processes, formation or selection of a mental representation of the task to act upon, allocation of mental resources to perform the task, monitoring of task performance, and evaluation of task performance (Kolligian & Sternberg, 1987).

The importance of metacognition to mathematical problem solving is well acknowledged in the literature (Hiebert & Carpenter, 1992). Metacognitive ability enables learners to adjust accordingly to varying task demands and contexts. Not only is metacognitive ability requisite for movement upward into more advanced and abstract stages, it also is a condition for operating at lower response levels. In other words, metacognition facilitates the selection and allocation of techniques and strategies for successful task completion. Whether metacognitive activity operates at the conscious or subconscious level depends on the nature and complexity of the problem to be solved, as well as the experiential and knowledge base of the problem solver. For example, during a think-aloud task, gifted students verbalized more metacognitive strategies as the mathematical problems increased in difficulty (Montague & Applegate, 1993b). In contrast, students with learning disabilities verbalized fewer strategies on the more difficult problems, suggesting cognitive overload and a shutdown of their ability to verbalize processes and strategies. For average achievers, the cognitive demands were sufficient to limit the metacognitive but not the cognitive strategies.

Metacognitive deficits appear to adversely affect the development and use of effective strategies for representing problems and executing solutions; as a result, these deficits impede progress in academic tasks requiring considerable strategic activity, such as mathematical problem solving. Because metacognitive processes are incremental (Case, 1985; Flavell, 1985), meaningful strategy instruction can occur only when the learner's cognitive maturity and reasoning ability are at a point of developmental readiness that allows interpretation and adaptation of appropriate learning strategies.

Two studies that are described in greater detail later in this chapter support this notion. In a descriptive study, Montague and Applegate (1993a) found signification differences on reasoning test scores and word-problem test scores between sixth-grade students and seventh- and eighth-grade students, collapsed across ability levels (gifted, average achieving, and learning disabled). In an intervention study designed to teach a cognitive–metacognitive strategy for mathematical problem solving to six students with learning disabilities, two sixth-grade students performed less well on word-problem tests following strategy instruction than seventh- and eighth-grade students and also did not meet the criterion for improved performance (Montague, 1992). The authors concluded that upper elementary school students may not be developmentally ready for instructional programs that emphasize higher level, abstract mathematical reasoning, nor will they profit from complex, comprehensive strategy instruction for mathematical problem solving. Thus, individual differences and the developmental nature of cognitive processing are important considerations when deciding for whom and under what conditions cognitive strategy instruction will be beneficial.

STUDENTS' COGNITIVE CHARACTERISTICS

Using their natural mathematical abilities, young children begin to acquire mathematical concepts early, through observation and spontaneous interactions with the environment (Ginsburg, chapter 2 of this book; Ginsburg & Russell, 1981). As they gain conceptual understanding, they easily acquire declarative or factual knowledge, through both informal and formal learning experiences. Children then use this declarative knowledge as they learn algorithms and become proficient in computational procedures. As children further engage in experiential learning, real-world problem solving, and formal school learning, they draw on their cumulative conceptual, declarative, and procedural knowledge to construct a fourth type of knowledge, strategic knowledge. This type of knowledge is demonstrated by an individual's ability to describe or apply mnemonic or problem-solving strategies that are either domain-specific or universal in nature.

Strategic knowledge, which is crucial to proficient problem solving, is acquired naturally by most children as they are exposed to problem solving (Siegler & Jenkins, 1989). These children need only the opportunity to solve problems as they construct and generalize knowledge of problem-solving strategies and become more effective, efficient problem solvers. Other children, however, may manifest disabilities that interfere with acquisition of knowledge, thus affecting school performance and general problem-solving abilities.

The acquisition of conceptual, declarative, procedural, and strategic knowledge in mathematics may be negatively affected by developmental delays, representational and retrieval problems, or visuospatial deficits (Geary, 1993). Developmental delays in the acquisition of conceptual knowledge seem to promote the use of developmentally immature arithmetical procedures and high procedural error rates. Difficulties in representation and retrieval of arithmetic facts from long-term semantic memory can hinder the acquisition of declarative knowledge. Visuospatial deficits (i.e., difficulties with spatial representation of numerical information) can preclude conceptual, declarative, and procedural understanding as a result of disorganized schemata and faulty mental representations of mathematical patterns. Additionally, deficits in strategy formation, selection, or allocation can adversely affect mathematical problem-solving performance (Montague & Applegate, 1993a; Swanson, 1990).

In his investigations of mnemonic strategies among normally achieving students and students with learning disabilities, Swanson (1990) found that qualitative differences in strategic behavior, rather than strategic deficiencies, explained how students with learning disabilities process information and solve problems. In our studies on mathematical problem solving, we found that the *quantity* of strategies used by middle school students does not vary much among ability groups. However, when we investigated differences in specific *types* of strategies in our attempt to understand why students with learning disabilities were poor problem solvers compared with their nondisabled peers, both quantitative and qualitative differences in strategy use emerged.

Interestingly, although the overall number of strategies did not differ among ability groups, we found that students with learning disabilities reported far more solution strategies, such as rereading problems and shifting computations, than the other students. The students with learning disabilities showed considerable difficulty in transforming linguistic and numerical information in word problems into appropriate mathematical equations and operations, and, as a result, they usually resorted to ineffective trial-and-error solution strategies and frequently performed a series of irrelevant computations. The most salient differences in strategic behavior were associated with problem-representation strategies (Montague & Applegate, 1993a, 1993b; Montague, Bos, & Doucette, 1991). When taught specific strategies for representing problems, such as paraphrasing or visualizing problems, their mathematical problem-solving improved (Hutchinson, 1993; Montague, Applegate, & Marquard, 1993; Zawaiza & Gerber, 1993).

Children who have mathematical disabilities require more than textbook or real-world practice in solving mathematical problems to become good problem solvers. These children generally need explicit instruction in problem-solving strategies as well as guided learning experiences in mathematical problem solving (Hutchinson, 1993; Montague, Applegate, & Marquard, 1993; Swanson, 1990). Additionally, research suggests that efficacy and attributional training enhances strategy-based performance for students with learning disabilities (Groteluschen, Borkowski, & Hale, 1990). Efficacy training promotes self-regulation by teaching students how to evaluate the effectiveness of strategies and revise and change strategies based on their efficacy. Attributional training focuses on the role of effort in performance and on using success-oriented dialogues and dialogues aimed at coping with failure; such dialogues link attributions to behavior (Reid &

Borkowski, 1987). Researchers have stressed the importance of expanding strategy instruction "to include executive processes that direct strategy use as well as attributional processes that energize strategic routines" (Groteluschen et al., 1990, p. 81).

COGNITIVE STRATEGY INSTRUCTION RESEARCH

Cognitive strategy instruction, which directly addresses students' comprehension and problem-solving deficiencies, has improved the academic performance, strategic knowledge, and affective responses of students with learning disabilities across academic and social–emotional domains (Englert, 1993; Graham & Harris, 1994; Montague, Marquard, & LeBlanc, 1993; Scruggs & Mastropieri, 1993; Scruggs & Wong, 1990; Vauras, Lehtinen, Olkinuora, & Salonen, 1993; Wong, 1992, 1993). Wong (1993) cited the following three reasons for why cognitive strategy instruction has moved from instruction in general strategies, which was typical in the 1970s and early 1980s, to instruction in domain-specific strategies: (a) Students have difficulty in relating general strategies to real-life tasks, (b) students are more motivated to learn and use strategies if they perceive the link between the strategy and the academic task, and (c) students who have domain knowledge learn general problem-solving strategies more easily. Although general problem-solving strategies such as association or mental imagery can be applied in many different situations, they seem to be acquired and applied better when grounded in specific content domains (Alexander & Judy, 1988; Pogrow, 1992).

Intervention research focusing specifically on strategy instruction for mathematical problem solving has shown generally positive results for upper elementary, middle school, secondary, and postsecondary students (Bennett, 1982; Case, Harris, & Graham, 1992; Huntington, 1994; Hutchinson, 1993; Marzola, 1987; Montague, 1992; Montague & Applegate, 1993a; Montague & Bos, 1986; Nuzum, 1983; Smith & Alley, 1981; Zawaiza & Gerber, 1993; Zentall & Ferkis, 1993). Although these studies vary somewhat with respect to specific strategies taught and instructional routines used, they usually include techniques for teaching students to represent problems by drawing a picture, constructtng a chart or table, or imagining the salient features of a problem.

Students who lack problem-solving strategies generally need explicit instruction in specific cognitive strategies (e.g., visualization, verbal rehearsal, paraphrasing, summarizing, estimating) to facilitate their reading, understanding, executing, and evaluating of problems (Wong, 1992). In contrast, students who have a repertoire of problem-solving strategies but use them inefficiently or ineffectively may need metacognitive strategies (e.g., self-instruction, self-monitoring, self-evaluation) to help them activate, select, and monitor strategy use (Graham & Harris, 1994). The content and duration of instruction vary, depending on its purpose. On the basis of intervention research in this area, several instructional principles, including cognitive modeling, verbal rehearsal, guided practice, corrective and positive feedback, and mastery learning, seem essential for strategy acquisition and application. The following application of cognitive strategy instruction illustrates how these instructional principles are used to facilitate students' strategy acquisition and application.

STRATEGY INSTRUCTION IN MATHEMATICS

Three intervention studies provide the foundation for the following illustration of cognitive strategy instruction (Montague, 1992; Montague, Applegate, & Marquard, 1993; Montague & Bos, 1986). These studies demonstrated the effectiveness of cognitive strategy instruction for improving the mathematical problem-solving performance of middle and secondary school students with learning disabilities. The students in these studies had been identified by their school districts as learning disabled. Additional criteria for participation included average intellectual ability, evidence of poor mathematical problem solving, and adequate reading comprehension and computational skills. In two of the studies (Montague, 1992; Montague & Bos, 1986), the students were instructed individually; in the third study (Montague, Applegate & Marquard, 1993), they were instructed in groups of 8 to 12 students.

The goal of instruction was to teach students a comprehensive cognitive and metacognitive strategy to help them solve one-, two-, and three-step mathematical word problems. The mathematical problem-solving model advocated here is derived from research in general problem solving (e.g., expert–novice studies, computer-simulation studies); mathematical problem solving; metacognition; and affective variables associated with problem-solving performance (see Montague & Applegate, 1993a, and Montague, Applegate, & Marquard, 1993, for brief reviews of this literature). Although the specific cognitive and metacognitive processes and strategies proposed in this model (see Figure 9.1) need further validation, research has suggested that successful problem solvers know these strategies and use them effectively (Montague & Applegate, 1993a, 1993b). Furthermore, when students

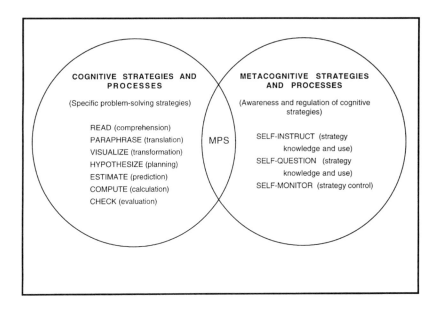

Figure 9.1. Cognitive–metacognitive model of mathematical problem solving (MPS).

with learning disabilities were taught to use these strategies, their problem solving improved to the level of their nondisabled peers' (Montague, Applegate, & Marquard, 1993).

The instructional program has three components: (a) assessing student performance and identifying students for whom the instructional program is appropriate; (b) explicitly instructing students in the acquisition and application of strategies for mathematical problem solving; and (c) evaluating student outcomes, with an emphasis on strategy maintenance and generalization. The components of the instructional program are described in the following sections, and recommendations for their implementation are provided.

Assessing Student Performance

Assessing student performance is an important precursor to enrollment in cognitive strategy instruction for several reasons. First, students may not have the skills in reading and writing, or, in this case, reading and mathematical computation, that are necessary for success in the program. Second, they may not be developmentally ready for the type of strategy instruction provided, particularly if the strategy package is complex and comprehensive.

Third, students' individual characteristics as well as their strengths and weaknesses with respect to strategic activity need to be ascertained prior to instruction. In this way, teachers and students can work together to develop goals within the context of the instructional program, which can then be tailored to meet the needs of individual students. Periodic assessment at regular intervals during instruction is also important for determining students' progress as they learn and practice strategies. Additionally, systematic follow-up assessment must be built into the overall program to determine whether students are maintaining and generalizing strategies.

Various types of assessment are recommended to provide the information needed to place students and develop individualized programs. Standardized measures such as the Wide Range Achievement Test–Revised (Jastak & Wilkinson, 1984) or specific subtests of the Woodcock-Johnson Psycho-Educational Battery (Woodcock & Johnson, 1989) can be used to assess a student's reading or computational abilities relatively quickly. The instructional program for mathematical problem solving described here presumes a reading grade equivalent of approximately 3.5; demonstrated skill in adding, subtracting, multiplying, and dividing whole numbers and decimals; and developmental readiness in metacognition and reasoning.

According to neo-Piagetian theory, metacognitive abilities reach maturity as children approach the formal operational stage of development, or somewhere around 12 to 14 years of age (Flavell, 1985). For this reason, students younger than 12 or students whose cognitive development is delayed or different may not gain much from complex, comprehensive cognitive strategy instruction. Rather, systematic exposure to problems in a variety of contexts, modified and simplified strategy instruction, and the opportunity to solve problems using concrete materials and examples may be more appropriate for their developmental level.

To gain insight into how students process information and solve problems, qualitative approaches to assessment have been recommended (Bryant & Rivera, chapter 5 of this book; Ginsburg, Jacobs, & Lopez, 1993; Montague, 1992). These include structured and clinical interviews, thinking-aloud exercises, self-report procedures, and stimulated recall activities. To assess students' attitudes toward mathematics and their knowledge, use, and control of mathematical problem-solving strategies, the Mathematical Problem Solving Assessment-Short Form (MPSA-SF; Montague, 1992) is recommended. This instrument is an abridged version of an extensive cognitive–metacognitive interview for mathematical problem solving (Montague & Applegate, 1993a; Montague, Applegate, & Marquard, 1993; Montague & Bos, 1990). The MPSA-SF includes 3 word problems, 5 Likert-type items, and 35 open-ended questions, and takes about 25 minutes to adminster and score. Individual cognitive profiles provide informal information about students' perceptions of their mathematical ability, their attitude toward mathematics and mathematical problem solving, and their knowledge, use, and control of problem-solving strategies. Appendix A contains the word problems and selected interview questions from the MPSA-SF. Students' cognitive profiles are based on their responses to the 40 items.

Figure 9.2 shows a completed profile for Armando, an eighth-grade student who demonstrated very little strategic knowledge, which may explain his inability to apply strategies when he solved the more difficult two- and three-step word problems. Armando showed relative strengths in computation, reading, hypothesizing, and relating mathematical problems to real-life experiences. Recommendations for Armando included providing comprehensive strategy instruction, using think-aloud as a technique to help him "think through" the problem, and capitalizing on his strength in relating problems to real life by having him "act out" multiple-step problems.

Other students may show different strengths and weaknesses in strategy knowledge, use, and control. For example, a ninth grader solved two of the problems correctly. Although he predicted the correct operations for the third problem, he did not represent it appropriately. His profile indicated that he had limited knowledge of problem representation strategies and few strategies for checking his work and monitoring his performance. Recommendations for this boy included teaching problem representation and self-monitoring strategies while reinforcing the self-management strategies he already used.

Analyses of student performance help teachers plan mathematics lessons that directly address individual student attributes. Even more important than assessing students' problem-solving accuracy is assessing their strategic performance. Assessment of problem-solving performance should also include an evaluation of attitude, which seems to play an important role in academic performance (Wong, 1994): When students have a positive perception of their ability and a good attitude toward mathematics, they most likely will approach instruction in a positive manner and will make a commitment to learning. These factors are important in effective strategy instruction (Ellis, 1993).

INSTRUCTING STUDENTS

The instructional program described in this section was developed over a 10-year period specifically to investigate the effects of cognitive strategy instruction on students' mathematical problem-solving performance. Three

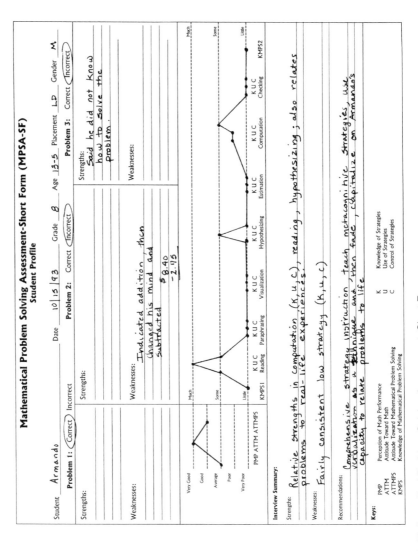

Figure 9.2. Mathematical Problem-Solving Assessment–Short Form.

intervention studies were conducted. In the first study (Montague & Bos, 1986), six secondary school students with learning disabilities were taught a comprehensive cognitive strategy consisting of eight cognitive processes. Teaching procedures for strategy acquisition and application combined direct instruction and cognitive–behavioral procedures, including modeling, rehearsal, feedback, and mastery tests. Results indicated substantial improvement on two-step word problems for all students following intervention, evidence of generalization to three-step problems, and maintenance of performance levels 3 months later.

The cognitive strategy was refined, and metacognitive components were added for the second study (see Figure 9.1), which was conducted with six middle school students with learning disabilities (Montague, 1992). One of the purposes of that study was to investigate the differential effects of the cognitive and metacognitive components of instruction. Although the study was limited in its ability to identify the elements of the instructional package that were most crucial to student improvement, it did suggest that the combination of cognitive and metacognitive strategies was more effective than either cognitive or metacognitive strategies taught in isolation.

In the third study, 72 junior high school students with learning disabilities participated (Montague, Applegate, & Marquard, 1993). Instruction was provided to groups ranging from 8 to 12 students in separate classrooms during their regularly scheduled 50-minute mathematics classes. This study also attempted to investigate the effects of cognitive strategies versus metacognitive strategies versus a combination of strategies on students' performance on one-, two-, and three-step word problems. Statistically, the conditions did not differ significantly on the various trials of word-problem solving; however, there were subtle indications that the combination of cognitive and metacognitive strategies was the optimal learning package. First, a trend toward more rapid progress was evident for the group that received the combination of cognitive and metacognitive strategies as their instructional package, and, second, this group seemed more likely to maintain improved performance.

The study lasted 14 days (including 2 days of posttesting). Maintenance probes were taken at 3-, 5-, and 7-week intervals, with the final maintenance test preceded by a booster session consisting of review and practice. All lessons were scripted to enhance treatment fidelity. The instructional routine for strategy acquisition included discussion of current performance, description of the cognitive processes and metacognitive activities associated with each process (see Figure 9.3), and verbal rehearsal of the processes and activities. Wall charts and booklets for home study were used during this period. Students were required to memorize the seven processes to 100% criterion

Read (for understanding)
Say: Read the problem. If I don't understand, read it again.
Ask: Have I read and understood the problem?
Check: For understanding as I solve the problem.

Paraphrase (your own words)
Say: Underline the important information. Put the problem in my own words.
Ask: Have I underlined the important information? What is the question? What
            am I looking for?
Check: That the information goes with the question.

Visualize (a picture of a diagram)
Say: Make a drawing or a diagram.
Ask: Does the picture fit the problem?
Check: The picture against the problem information.

Hypothesize (a plan to solve the problem)
Say: Decide how many steps and operations are needed. Write the operation
            symbols (+ − × ÷)
Ask: If I do —, what will I get? If I do —, then what do I need to do next? How
            many steps are needed?
Check: That the plan makes sense.

Estimate (predict the answer)
Say: Round the numbers, do the problem in my head, and write the estimate.
Ask: Did I round up and down? Did I write the estimate?
Check: That I used the important information.

Compute (do the arithmetic)
Say: Do the operations in the right order.
Ask: How does my answer compare with my estimate? Does my answer make
            sense? Are the decimals or money signs in the right places?
Check: that all the operations were done in the right order.

Check (make sure everything is right)
Say: Check the computation.
Ask: Have I checked every step? Have I checked the computation? Is my
            answer right?
Check: That everything is right. If not, go back. Then ask for help if I need it.

Figure 9.3. A cognitive–metacognitive strategy for mathematical problem solving.

and paraphrase the metacognitive activities. The teacher (a) modeled strategy recitation and application using an overhead projector while solving problems, (b) exchanged roles with students during demonstration activities, and (c) gave positive corrective feedback during guided practice sessions. Instructional procedures for strategy application included independent practice, practice in pairs, teacher–student role exchange during demonstration exercises, and explanations of problem solving by students using the chalk-board or overhead projector. Appendix B contains the teaching script for the first day of instruction.

Evaluating Student Outcomes

The following questions should be asked separately when evaluating the effects of cognitive strategy instruction on student learning: (a) Did the student *acquire* the strategy? (b) Can the student *apply* the strategy? (c) Did the student *maintain* the strategy? and (d) Did the student *generalize* the strategy? Thus, we must evaluate cognitive strategy instruction with respect to four levels of learning: strategy acquisition, application, maintenance, and generalization.

Research has established that if students with learning disabilities are developmentally ready and meet other established criteria for a specific domain (e.g., a minimum reading grade equivalent, basic writing skills), and if the instruction is sound and delivered well, then they usually can acquire strategies without much difficulty (Wong, 1994). Furthermore, if students know the strategy and are given appropriate tasks and enough practice using the strategy in the designated domain (e.g., mathematical problem solving, narrative composition, expository reading), then they usually can apply the strategy appropriately (Pressley & Associates, 1990).

Progress is being made with procedures to promote maintenance of learned strategies, as well. For example, in the group intervention study on mathematical problem solving, we conducted three maintenance tests to determine whether students were maintaining performance levels on word-problem tests (Montague, Applegate, & Marquard, 1993). We were interested in how long the students would be able to maintain the strategy following instruction; we expected them to decline in performance over time. On the first maintenance measure, administered 3 weeks following instruction, the students met the preset criterion of at least 7 correct on a 10-problem test. On the second maintenance measure, administered 5 weeks after the first one, their average score dropped below this criterion. We assumed that their performance would continue to decline if no further intervention were provided. So, prior to the third maintenance measure, administered 7 weeks later, we provided 1 day of strategy acquisition review (see Appendix C for the script of this lesson) and 1 day of guided practice using the same procedures we used during strategy application practice. Only the students who received the combination of cognitive and metacognitive strategies as their instructional package met the preset criterion for mastery. We concluded that distributed practice and periodic booster sessions may be enough to maintain strategy use for students whose performance deteriorates over time.

Unfortunately, intervention researchers have not been as successful in teaching students to generalize strategies as they have been in teaching strategies initially, in teaching domain-specific application, and, recently, in promoting strategy maintenance. Wong (1994) addressed strategy generalization in her discussion of strategy transfer, which she defined as "students' spontaneous, unprompted, appropriate use of previously learned strategies in tasks/situations that differ from those in which the strategies were originally learned" (p. 110). She suggested two instructional parameters to promote transfer of learned strategies by students with learning disabilities. First, she underscored the importance of mediating student "mindfulness" during strategy learning and transfer. Salomon and Globerson's (cited in Wong, 1994) definition of mindfulness is a "state of mind involving volitional, metacognitively guided employment of nonautomatic, usually effort demanding processes" (p. 111). Mindfulness allows individuals to confront new tasks or situations in a confident manner and use their strategy knowledge to select appropriate strategies. Wong's suggestions for inculcating mindfulness in students include (a) providing students with ample instructional scaffolding with systematic modeling of active thinking, (b) encouraging interactive dialogues with peers, (c) scheduling programmed practice with transfer tasks, and (d) asking students to verbalize their rationale for selecting and using particular strategies. Strategy generalization is the ultimate test or cognitive strategy instruction and must be addressed in future intervention research studies.

To facilitate evaluation of these four levels of strategy instruction, the following questions can be asked:

1. Strategy Acquisition
- Can the student recite the strategy from memory and/or paraphrase the strategy (depending on the goal)?
- Can the student explain or define the terms of the strategy?
- Does the student understand the rationale for learning the strategy?

2. Strategy Application
- How does the student's performance on domain-specific tasks (e.g., mathematical word problems, paraphrasing passages) following instruction compare with his or her performance prior to instruction (posttest vs. baseline)?

- Has the student reached mastery by achieving a certain preset criterion for acceptable performance?
- Can the student use the strategy to detect and correct errors (self-monitor performance)?

3. Strategy Maintenance
- Did the student maintain performance levels on domain-specific tasks over time?
- Can the student explain how the strategy was used to solve the problem or complete the task?
- If the student did not maintain performance levels, did a booster session improve performance to mastery level?

4. Strategy Generalization
- Did the student use the strategy successfully in other tasks?
- Did the student use the strategy successfully in other situations or settings?
- Can the student verbalize his or her rationale for selecting and using the strategy?

CONCLUSIONS

Cognitive strategy instruction is a promising alternative to current approaches for teaching mathematics to students with learning disabilities. For students who have learned basic mathematical skills but cannot apply them successfully when solving mathematical problems, cognitive strategy instruction should be a part of their curriculum. Effective instruction, however, depends on the teacher's understanding of the theoretical foundations for strategy instruction as well as the efficacy research. Furthermore, teachers need to be cognizant of the notion that strategy instruction may not be appropriate for all students, and that instruction must be tailored to individual students. Additionally, if we expect students to generalize the strategies they have learned, they must become actively involved in the learning process by participating in goal setting and evaluating their own performance. Cognitive strategy instruction for mathematical problem solving, as described in this article, can be effective at improving the mathematical performance of students with learning disabilities.

Appendix A: Selected Items From the Mathematical Problem Solving Assessment–Short Form (MPSA-SF)

Part A

Examiner: *"Here are three math word problems. I will read them to you. You do not need to solve them. (Read the problems.) Now I would like you to answer the following questions. I will write your answers."*

1.   Bill and Shirley need to arrange the chairs for a play that the class is having. They took 252 chairs from the storeroom. Their teacher told them to make rows of 12 chairs each. How many rows will they have?

2.   Four friends have decided they want to go to the movies on Saturday. Tickets are $2.75 for students. Altogether they have $8.40. How much more do they need?

3.   Chain sells for $1.23 a foot. How much will Farmer Jones have to spend for chain in order to enclose a 70-foot by 30-foot patch of ground, leaving a 4-foot entrance in the middle of each of the 30-foot sides?

	Very Poor	Poor	Average	Good	Very Good
Desribe your math skills.	1	2	3	4	5
Describe your math grades.	1	2	3	4	5
Describe how well you solve math word problems.	1	2	3	4	5

Do you like math?

	Not at all	1/4 of the time	1/2 of the time	3/4 of the time	Always

Tell me what you remember being taught about how to solve math word problems.

A strategy is a general plan or a specific activity people use to solve problems. Tell me about any strategies you use to solve math word problems.

How do you read math word problems?

As you read, how do you help yourself understand the problem?

Do you put what you read into your own words?

How do you do this? Now I would like you to put Problem 3 into your own words. (Give the student the problems.)

Do you ever make a drawing of the problem or see a picture of the problem in your mind? (Have the student clarify responses. If necessary, use the following probes.)

Probes:
What kind of picture? How often do you use drawings or pictures? When do you make drawings of problems? Under what conditions do you make drawings or see pictures in your mind? Which problems? (Have student draw a picture of one of the problems.)

How do you make a plan to solve a math word problem?

How do you know which operations to use (such as adding, subtracting, multiplying, and dividing)?

How do you decide how many steps are needed to solve a math story problem?

Estimation is making a prediction about the answer using the information in the problem. How does estimation help in solving math word problems?

What do you do when you compute answers to word problems?

How do you check that you have correctly completed a math word problem?

Appendix B: Cognitive Strategy Instruction for Mathematical Problem Solving—Teaching Script for Day 1

Teacher: For the next 7 days I am going to teach you to use a strategy for solving math word problems. Many of you told me that you are not very good at solving word problems and that you do not like to solve them. You also told me that you think it is important to be a good math problem solver and that you would like to improve your problem-solving skills. Why do you want to improve your math problem solving?

Student: Better grades, important skill, etc.

Teacher: You are right. *(Good grades, managing money, you need it on the job, etc.)* are important reasons for becoming a good math problem solver. You all have some good math skills. I will teach you how to use those math skills when you solve math word problems. Let's look at the results of the two problem-solving tests you have taken. *(Distribute the folders and explain the baseline scores and how to read the graphs.)* The first point tells how you did on the 6 problems you solved during the interview. The second point tells how you did on the 10 problems you solved in class the other day. Some of you did fairly well, and other people did fairly poorly. We would like everyone to do well on all the problems all of the time. Your goal will be to get at least 7 problems correct on each word-problem check that you do. If you work hard in class during the next 3 weeks, you will improve your problem solving and your scores. Questions? Also, many of you told me that you do not like solving word problems. I can understand this because most people do not like to do things that they do not do well. When you become better math problem solvers, you will like problem solving more than you do now. The goal of this project is to have you become better math problem solvers. What is your goal? Questions? *(Collect the folders.)*

First, I will teach you a seven-part strategy for solving math problems. You will then practice using the strategy. This will require 7 days. Then you will take a word-problem check. After that you will practice using the strategy to solve more math word problems. This will take 5 days. Then you will take another word-problem check. That will conclude the instruction, and your regular teacher will return. For the next few weeks after that, I will interview each one of you individually and give you more word-problem checks. The study will be over on April 25. Questions?

All right. Let's begin.

People who are good math problem solvers do several things in their head when they solve problems. They use several processes. Raise your hand if you know what a process is. *(Call on students.)* A process is a thinking skill. What is a process?

(Use the Direct Instruction [DI] procedure : First, ask a question. Then, have the students answer in unison. Model the response. Then ask the same question and call on students individually for responses. For example, ask, "What is a process? The response should be, "A process is a thinking skill.")

Good problem solvers tell us they use the following seven processes when they solve math word problems. I have written these processes in a booklet for you to keep and study at home and also on a big chart to use in class while you are learning the strategy.

(Show the chart with only the names of the cognitive processes to the students. Point to each process and read the process name, explain its meaning, and ask questions.)

First, good problem solvers *read* the problem for understanding. *(Point to the process name.)*

Why do you read math word problems? *(Use the DI procedure.)* "I read for understanding."

Then, good problem solvers *paraphrase* the problem. Raise your hand if you know what "paraphrase" means. *(Call on students.)*

Paraphrase means to put the problem into your own words and remember the information.

What does paraphrase mean? (DI)

The third process is to *visualize.* Raise your hand if you know what "visualize" means. *(Call on students.)* When people visualize word problems, they use objects to show the problem, or they draw a picture or a diagram of the problem on paper, or they make a picture in their head.

How do people visualize? (DI)

Next, good problem solvers *hypothesize.* Raise your hand if you know what "hypothesize" means. *(Call on students.)*

Hypothesize means to set up a plan to solve the problem.

What does hypothesize mean? (DI)

Then people *estimate* the answer. Raise your hand if you know what "estimate" means. *(Call on students.)*

To estimate means to make a good prediction or have a good idea about what the answer might be using the information in the problem. Raise your hand if you know what a prediction is. *(Call on students.)* People estimate or predict answers before they do the arithmetic. After they do the arithmetic and get the actual answer to the word problem, they compare their actual answer with the estimated answer. This helps them to decide if the answer they got is right or if it is too big or too small.

What is estimating? (DI)

So, after good problem solvers estimate their answer, they do the arithmetic. We call this *computing.*

What is computing? (DI)

Finally, good problem solvers *check* to make sure that they have done everything right. That is, they check that they have used the right operations and completed all the necessary steps, and that their arithmetic is correct. To check computation, people sometimes use addition to check subtraction problems and use multiplication to check division problems.

Why do you check math word problems? (DI)

All right, here are the seven processes and the explanations for each one. *(Review the chart listing the processes.)*

People who are good math problem solvers also do several things in their head when they solve problems. First, they SAY different things to tell themselves what to do. Second, they ASK themselves questions. Third. they CHECK to see that they have done what they needed to do to solve the math problems. I have put each SAY, ASK, and CHECK activity with the right process on these charts. I have these activities written in these little booklets for you to keep and study. I will give them to you at the end of class today for you to take home and study. I also wrote the activities on this big chart for you while you are learning the activities to solve word problems. We will call these processes, and we will call the activities that go with them, "strategies." Now I am going to read the entire strategy through once. Then we will read it as a group. Then I will call on each one of you to read the strategy.

(Show booklets and charts to students. Point to each activity and verbalization as you read and explain it.)

All right, now I would like you to read through the charts. I will help you with the words if you need help. *(Group reading—twice)*

Now I would like you to read the process and the words *SAY, ASK,* and *CHECK,* and I will read the activities. *(Group)*

Now I will read the process and the words *SAY, ASK,* and *CHECK,* and you will read the activities. *(Group)*

Now I want you to read everything. *(Individuals—one time each)*

Tonight at home I want you to memorize the seven processes and what each one means. You do not need to memorize the activities, although I want you to know them. You can remember the processes by using the first letter of each process: RPV-HECC. Here are your study cards.

Appendix C: Booster Session for Cognitive Strategy Maintenance

Day 1:   **Strategy Review**
Day 2:   **Practice**
Day 3:   **Maintenance Test**

Day 1: Strategy Review

Teacher: You all have done remarkably well in this study on mathematical problem solving. I am very proud of you and happy with the progress you have made. Although most of you did fairly well on the last word-problem check, it is important that you continue to do well and solve problems using the mathematical problem-solving strategies you have learned. So, for the next 3 days you will practice using strategies. In this way, you will be less likely to forget math problem-solving strategies and more likely to use them when you solve mathematical word problems.

First, let's review the strategy.

(Have all the students recite the strategy from memory, providing cues. When needed, use the small chart to cue students. Repeat until all the students recite the strategy to 100% criterion.)

(Then, review the metacognitive activities for each strategy. Have students try to recall from memory the activities for each component of the strategy. Use the transparencies and the small cards to compare students' responses to the actual activities.)

Teacher: Remember, as you become better strategy users and also better mathematical problem solvers, you will begin to use only parts of the strategy and only when you need them. Some people will use all parts of the strategy all of the time. Everyone is different. How did you use the strategy once you learned it?

(Elicit from students various ways they adapted the strategy. Then, discuss how strategy adaptation to individual styles is important. Also reinforce the idea that using the

entire comprehensive strategy may be important at times. Reinforce the notion that strategy components can be used recursively; that is, they can be used in a different sequence, or be used several times, or be varied slightly when solving a problem. Show examples.)

Demonstration and Review

Problem 1:   Demonstrate using the strategy while solving a problem.

Problem 2:   Have students tell you what to do as you solve Problem 2. Call on students individually. Tell them they can adapt the strategy by indicating what they would do.

Problem 3:   Have students solve problems independently. Then, select a student to demonstrate solving the problem and, again, discuss how other students used different ways to solve the problem.

Problem 4:   Have students tell you what to do to solve Problem 4. Call on students individually.

Then, practice strategy recitation from memory until the class ends.

10. A Life Skills Approach to Mathematics Instruction: Preparing Students with Learning Disabilities for the Real-Life Math Demands of Adulthood

JAMES R. PATTON, MARY E. CRONIN, DIANE S. BASSETT,
AND ANNIE E. KOPPEL

The teaching of mathematics is an integral part of the curriculum at both the elementary and the secondary level. As other authors of chapters in this book (Ginsburg; Miller & Mercer) have pointed out, mathematics can be a difficult subject area for some students with learning disabilities.

Some degree of mathematics proficiency is needed for most jobs; for some jobs, it is the essence of what the worker does (e.g., accountant, cashier). In reality, mathematics pervades almost everything we do as adults. Most of the competence we need, however, does not involve complex, advanced mathematical understanding. Rather, it entails the application of basic arithmetic operations to everyday situations.

Given that only 18.1% of youth with learning disabilities go on to 2-year or 4-year colleges (Wagner, Blackorby, Cameto, Hebbeler, & Newman, 1993)—settings in which one may certainly need higher levels of math competence—it is prudent and necessary to look closely at the noncareer math skills that are needed for survival in everyday living. Those students with learning disabilities for whom higher education is possible must be exposed to higher levels of mathematics while they are in high school, as such exposure is necessary for dealing with the math demands of their college course work. However, attention must be given to teaching the math skills that will be used on the job, at home, and in the community, as a matter of routine for all students with learning disabilities, and especially for non–college-bound students, who constitute the majority of students with learning disabilities.

An index of the use of various math-related life skills by an individual with learning disabilities is available from the findings of the National Longitudinal Transition Study (Wagner et al., 1991). One area about which data were collected involved the extent to which students performed various financial-management activities. Although 44.4% of the young adults with learning disabilities in the study who had been out of school for up to 2 years had savings accounts, very few of these individuals had checking accounts (8.1%), a credit card in their own name (8.1%), or other investments (.4%). These data suggest that students with learning disabilities either are not being prepared to take these financial avenues or are choosing not to pursue them. Regardless, all of these areas involve life skills that most adults will need.

Certain recurrent themes underlie the impetus to better prepare students for the realities of adulthood (Polloway, Patton, Epstein, & Smith, 1989). Some of the most important themes are as follows:

- The educational programs of many students with learning disabilities are not preparing them to meet their current and future needs.
- An overwhelming need exists to reexamine programming for students with learning disabilities at the secondary level and to consider more relevant, innovative options to prepare them for various postsecondary contexts.
- In addition to reexamining the secondary programs, professionals must carefully review the elementary and middle school programs of students with learning disabilities to determine how preparation for subsequent environments can be systematically included throughout the curriculum and the entire school experience.
- A significant number of students with learning disabilities are not finding their present school experience to be beneficial or relevant to their daily lives and thus are dropping out during the early years of high school.
- Students need continuing support services even after they systematically exit school—and, in fact, throughout adulthood.
- As noted earlier, only a relatively small percentage of students with learning disabilities go on to higher education and, even for those who do, life skill competence is essential.

This chapter is primarily about teaching the life skills math that nearly every student with learning disabilities will need to function successfully as an adult. It is organized into three major sections. The first section provides a foundation for teaching life skills—what they are and why it is important to include them in the curriculum—and identifies the common situations in which most adults need to display math competence. The second section relates the life skills instruction to the current reform movements affecting math education. The third section provides some suggestions for teaching life skills math to students with learning disabilities by examining various curricular and instructional dimensions.

NATURE OF LIFE SKILLS MATH

Life skills can be conceptualized as those specific skills or tasks that contribute to an individual's successful independent functioning, across a variety of situations. They are influenced by local and cultural contexts. Life skills competence is needed in a range of domains, including employment, further education, home and family, leisure, health, community involvement, interpersonal relationships, and personal development (Cronin & Patton, 1993). Mathematic competence is woven into all of these areas.

Math is a significant part of all of our lives, affecting successful functioning on the job, in school, at home, and in the community. Most individuals are able to generalize the math they learned in school to the wide variety of real-life situations that require math competence. However, for a significant number of students with learning-related problems, this transfer to everyday living remains elusive.

A case for the inclusion of life skills–oriented math content in programs for students with learning disabilities has already been made. The specific content to be included in any given program should be determined locally; however, some life skills math topics are generic (i.e., found in most settings). A list of contexts requiring math competence in select areas is provided in Table 10.1.

The following sections explore three conditions where life skills math competence is needed: in the workplace, in postsecondary training/education settings, and in everyday activities at home and in the community.

Workplace Math

Every job involves some use of mathematical concepts. Personal interest in job-related math usually focuses on one's financial compensation for

Table 10.1

Math Skills Typically Encountered in Adulthood

Life demand	Applied math skills					
	Money	Time	Capacity/volume	Length	Weight/mass	Temperature
Employment						
Transportation	x	x		x		
Pay						
• wages	x	x				
• deductions	x					
• taxes	x					
• retirement	x	x				
• investment	x	x				
• savings	x					
Commission						
• straight or graduated	x					
Hours worked		x				
Overtime	x	x				
Breaks/lunch	x	x				
Deadlines		x				
Further education						
Budgeting	x	x				
Costs	x					
Financing	x					
Time management						
• requisite course hours		x				
• scheduling		x				
• extra curricular		x				
• meetings		x				
Home/family						
Budgeting	x	x				

(table continues)

(Table 10.1 continued)

Life demand	Applied math skills					
	Money	Time	Capacity/volume	Length	Weight/mass	Temperature
Bills						
• payment options	×					
• day-to-day costs	×	×				
• long-term purchases	×					
Locating a home						
• rental or purchase	×	×	×	×		
• moving	×	×	×		×	
• insurance	×	×				
• contracts	×	×				
• affordability	×					
• utilities	×					
Mortgage	×	×				
Home repair/maintenance	×	×	×	×	×	×
Financial management						
• checking/savings account	×					
• ATM	×			×		
• credit cards	×	×				
• insurance	×	×				
• taxes	×	×				
• investment	×	×				
Individual/family scheduling		×		×		
Automobile						
• payments	×	×				
• maintenance	×	×	×	×	×	×
• repair	×			×		
• depreciation	×					
• fuel costs	×		×			
Thermostat		×				×
Cooking	×	×	×	×	×	×

(table continues)

(Table 10.1 continued)

	Applied math skills					
Life demand	**Money**	**Time**	**Capacity/volume**	**Length**	**Weight/mass**	**Temperature**
Yard maintenance	x	x	x	x	x	x
Home remodeling	x	x	x	x	x	x
Decorating	x	x	x	x	x	x
Shopping						
• comparing prices	x	x	x		x	
Laundry	x	x	x			x
Leisure pursuits						
Travel	x	x		x	x	x
Membership fee	x	x				
Subscription costs	x	x				
Reading newspaper	x	x	x	x	x	x
Equipment costs						
• rental or purchase	x	x				
Sports activities	x	x		x	x	x
Entertainment (e.g., movies, videos, performances, sporting events, cards, board games, electronic games)	x	x		x		x
Lottery	x					
Hobbies		x	x	x	x	
Personal responsibility and relationships						
Dating	x	x				
Scheduling		x				
Anniversaries/birthdays, etc.	x	x	x			
Correspondence	x	x			x	
Gifts	x	x				

(table continues)

(Table 10.1 continued)

Life demand	Applied math skills					
	Money	Time	Capacity/volume	Length	Weight/mass	Temperature
Health						
Physical development						
• weight					X	
• height				X		
• caloric intake	X	X		X		
• nutrition	X	X			X	
Physical fitness program	X	X		X	X	
Doctor's visits	X	X		X	X	X
Medications	X	X	X			
Medically related procedures (e.g., blood pressure)		X				X
Community involvement						
Scheduling		X		X		
Voting		X		X		
Directions				X		
Public transportation	X	X				
Menu use	X	X				
Tipping	X					
Financial transactions						
• making/receiving change	X	X				
• fines/penalties	X	X				
Phone usage	X	X				
Using specific community services	X	X		X		
Emergency services	X	X		X		X
Civic responsibilities						
• voting		X		X		
• jury duty		X				

work performed. The actual job one holds will determine whether compensation is in the form of hourly wages, a monthly salary, commissions from sales, or some combination of these options. Regardless of how the gross amount on the paycheck is determined, workers have a keen interest in verifying the correctness of the amount of their check after deductions. This would require that they understand how net pay is determined and the reason for the various deductions (retirement, taxes, insurance, savings, etc.), and that they be aware of the amount of their deductions.

The type and complexity of math needed for specific job duties will vary with the nature of the job. Almost all jobs demand some math competence. For example, salespeople and cashiers need to know how to give correct change to buyers, total sales for the day, and balance the cash drawer with receipts. A person who works for a delivery service must be very skillful at knowing addresses, managing time, and maintaining a log—all of which require math.

The point is that certain general math skills are part of most jobs, and additional, vocation-specific math skills are also usually required, depending on the type of job. Preparation for both the generic skills and the more specific skills is critical to success. Generic math skills should be covered in the general curriculum; the career-specific math skills need to be refined via vocational preparation. One way in which many training programs are teaching various math skills is through integrative approaches within the context of vocational classes (Barbieri & Wircenski, 1990; Pickard, 1990).

Math in Postsecondary Training or Education

Any type of postsecondary education, from a couple of months of vocational training to several years of undergraduate/graduate education, will require various types of mathematical competence. The extent of needed proficiency will depend on the nature of the training or program of study.

Postsecondary vocational training programs are designed to prepare the individual for specific jobs. Within this training, the student will be exposed to different facets of the job that involve math performance. For example, an individual studying to become an auto mechanic will need to become skilled at recognizing and using different tools that are calibrated in either standard or metric units. The general education requirement of higher education programs typically has mathematic requirements, such as algebra courses. As a result, a sound foundation in mathematics can be beneficial to students as they pursue a given degree program. Even programs of study that on the surface do not seem to be math laden (e.g., social work) will involve some amount of applied math.

Everyone, regardless of postsecondary education setting, will need to deal with the everyday math discussed in the next section. Areas such as time management and personal finance are very much a part of the lives of young adults pursuing postsecondary training or education.

Everyday Math in Home and Community

Much of the math that we use on a daily basis involves in-home or community routines. Much of Table 10.1 highlights areas in which people will need to demonstrate math proficiency. Without question, some of the math must be worked on independently, for example, when paying for groceries or estimating how long it will take to run an errand. However, other everyday tasks involving math can be accomplished with the support of others (e.g., tax preparation, landscaping). Nevertheless, it seems logical that many of these types of topics would be covered comprehensively in the math curriculum.

An important point to remember is that much of the math that we use on a daily basis involves *estimation* rather than precise calculation. Certainly, precision is useful in maintaining a checkbook, but in many other areas, estimation skills are essential. For instance, scheduling one's day demands estimates of the time it will take to travel or how long a meeting will take. The number of grocery items one buys is guided by judging their estimated cost against available funds.

Another important point is that almost all of the situations identified in Table 10.1 require the individual to apply math skills in problem-solving situations. The more prepared a student is for the typical events of adulthood that require math, and the more proficient the student is at using the basic skills of arithmetic, the more likely it will be that he or she will solve the math-related challenges of adulthood. The educational challenge is to prepare students, within existing curricular structures, for real-life situations.

LIFE SKILLS MATH IN THE CONTEXT OF REFORM

One of the greatest challenges facing advocates of life skills instruction is demonstrating how functional math skills relate to the various reform initiatives. At first glance, a workable relationship may seem strained; however, life skills math and emerging standards for math curricula do not have to be perceived as separate tracks. Central to the document *Curriculum and Evaluation Standards for School Mathematics* (National Council of Teachers of

Mathematics [NCTM], 1989) is the belief that teachers must help students become better problem solvers through the use of interactive, hands-on mathematics instruction. Rote memorization and pencil-and-paper exercises are minimized in favor of "investigating and reasoning, means of communication, and notions of context. In addition, for each individual, mathematical power involves the development of personal self-confidence" (NCTM, 1989, p. 5).

The NCTM (1989) Standards include statements of what students should learn in Grades K through 4, Grades 5 through 8, and Grades 9 through 12. This includes areas such as Mathematics as Problem Solving, Mathematics as Communication, Mathematics as Reasoning, Mathematical Connections, Number Systems and Number Theory, and Computation and Estimation. In short, the NCTM Standards constitute a long-awaited reform effort that, if implemented appropriately, will guide students to higher levels of mathematics understanding, reasoning, and application. Although the NCTM content standards would appear to work harmoniously with life skills applications (Marzano & Kendall, 1995), a direct application of life skills instruction in the standards has not been formally explored.

According to certain professionals (i.e., Hofmeister, 1993; Hutchinson, 1993; Mercer, Harris, & Miller, 1993; Rivera, 1993), the NCTM Standards have failed to provide valid, data-based instructional programs for a diverse student population. Minimal reference is made to students who may be at risk for math failure, including students with disabilities, regardless of type or severity of disability. Hofmeister noted the "troubling lack of reference to research of any kind" (p. 12) to validate the Standards and argued that unless such research is conducted, validated, and replicated, students most vulnerable to failure will continue in that pattern. This downward spiral may be exacerbated for students with math disabilities who receive the bulk of their instruction in general education settings, where often very few real adaptations or accommodations are used (Rivera, 1993).

Aside from reform efforts aimed specifically at math, other general education reform initiatives do not address or articulate the needs of students with disabilities. Goals 2000, the Educate America Act, is currently being implemented by states across the nation as a foundation upon which to build instructional practices. Goals 2000 directly addresses the following mathematics competencies to be mastered by students:

- Students in Grades 4, 8, and 12 will have demonstrated competency over challenging subject matter, including English, mathematics, science, foreign languages, civics and government, economic, arts, history, and geography.
- U.S. students will be first in the world in mathematics and science achievement.

- Every adult will be literate and will possess the knowledge and skills to compete in a global economy.

This reform act is applicable to all students, including those with disabilities. Goals 2000 includes Opportunity to Learn (OTL) standards, which are specifically targeted at students vulnerable to repeated failure and provide these students with a better chance to master these content standards.

The inclusion of life skills instruction in relevant classroom practices and instruction is critical for meeting not only the standards of Goals 2000, but those of the NCTM as well. There are several compelling arguments for including life skills instruction across a K–12 curriculum, for both typical learners and learners with disabilities. Life skills instruction can address both the NCTM Standards and Goals 2000 in meaningful, relevant ways. All students should master the content standards to progress from grade to grade and to eventually graduate; direct application of life skills instruction can help foster student motivation and subsequent comprehension of mathematical concepts, computation, and application. Additionally, life skills instruction directly relates to specific goals delineated in the NCTM Standards, including problem solving, reasoning, connections, estimation, measurement, patterns, and functions. Instruction can be enhanced by the consistent integration of life skills topics into existing course content and by using community-based experiences.

In addition to providing relevance and opportunities for real-life applications, life skills instruction in math also bridges the gap between theory and practice. When instruction utilizes real-life situations, such as estimating the costs of going on a date or purchasing supplies for a hobby, it places math into the real world of students' lives. Theoretical constructs take on new life, which can then be generalized to other situations.

Life skills can help students at all grade levels to master math content in nontraditional ways. For example, a student with learning disabilities is likely to find meaningful a real-life math problem involving square feet and calculating how many gallons of paint are needed to cover the exterior of a home. The problem could also include estimation, hourly wage, taxes, budgeting time, and the cost of materials.

The Division of Career Development and Transition (DCDT) of the Council for Exceptional Children developed a position paper that strongly depicts life skills instruction as critical for all students with disabilities, regardless of academic ability or educational setting (Clark, Field, Patton, Brolin, & Sitlington, 1994). Real-life applications of math hold the key to reaching students traditionally at risk for failure in traditional mathematics education and should be included as a key component in the development of professional, national, or state reform standards.

CURRICULAR CONSIDERATIONS

It is crucial that for teachers to find ways to include life skills math topics in the curricula of students with learning disabilities. The overriding theme of all life skills instruction is problem solving, as daily life presents one problem to solve after another. This section offers some suggestions for teaching life skills math within the parameters of current educational practice.

Problem Solving

As has been stressed throughout this article, a significant amount of instructional time should be dedicated to real-life problems. As was emphasized more than 15 years ago in *Agenda for Action* (National Council of Teachers of Mathematics, 1980), and is clearly noted in more recent reform initiatives, mathematics curricula should be organized around problem solving. Figure 10.1 suggests that all the major areas of mathematics must lead to functional problem solving.

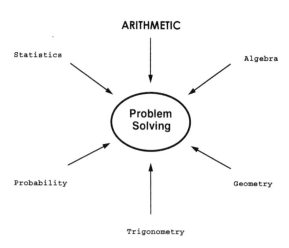

Figure 10.1. Relationship of areas of mathematics to problem solving.

Table 10.2

Functionality of Various Areas of Mathematics

Area of mathematics	Definition	Example of functional application
Arithmetic	Real numbers and computations involving them	(Most of Table 10.1)
Algebra	Generalization of arithmetic	Ratios—scale on map Cooking—adjusting recipes Income tax formulae
Geometry	Measurement, properties, and relationships among points, lines, angles, surfaces, and solids	Gift wrapping Map reading Wallpaper hanging
Trigonometry	Relationships between the sides and angles of triangles	Miniature golf Putting up a tent Interior design
Probability	Mathematical basis for prediction	Planning food for a party Card and board games Selecting which cashier's line to get into
Statistics	Collection, analysis, interpretation, and presentation of masses of numerical data	Price comparisons Baseball information Stock market Diet/nutrition Monitoring temperature of sick child

Many students with learning disabilities need to become proficient problem solvers in dealing with real-life situations. Most of the mathematics needed on a day-to-day basis involves the application of basic arithmetic principles to a variety of daily encounters—primarily measurement, as highlighted in Table 10.1. It is important to note that life skills mathematics is not restricted solely to arithmetic operations, as competence in other areas of mathematics may also be needed (see Table 10.2).

Changing Curricular Focus

Instructional attention to math-related life skills varies as a function of the level of schooling. At the elementary level, math instruction is focused on basic skill development, giving direct attention to or laying the foundation for the different areas of mathematics depicted in Figure 10.1. The challenge at this level is to relate math content to real-life situations as often as possible.

At the secondary level, math instruction typically is determined by the nature of one's program (college preparation, general education, vocational, or special). Life skills can be covered in a number of ways. Functional content can be integrated in courses that are not inherently life skills oriented. Additional life skills topics can enhance other courses that are already functionally oriented (e.g., consumer math). However, a significant number of students may not be exposed to this type of coursework. Furthermore, the quality of these courses will depend on the specific content covered (i.e., the textbook used) and the teacher's knowledge and skills.

Addressing Life Skills Content

Life skills can be taught in any type of educational arrangement; two basic requirements are a knowledge of what needs to be covered and the motivation/mandate to do so. The following sections highlight specific methods for incorporating life skills content into existing curricular structures or creating new coursework that is life skills focused.

Integration of Math-Related Life Skills into Existing Content. The overriding characteristic of this approach is the utilization of existing curricular content. This can be accomplished through either *augmentation* or *infusion* (Cronin & Patton, 1993). Both techniques use the already prescribed math scope and sequence by supplementing it with life skills content. Although both techniques require advanced planning, neither need become overwhelming to teachers who already have a great many demands on their time. Teachers are likely to find that, with the minimal preparation time involved and the availability of useful resources, integrating life skills content into math classes is beneficial to them and their students. Teachers often become more enthusiastic about teaching traditional topics, and students respond to the newfound relevance of these same topics to their lives.

An arbitrary and fine distinction exists between augmentation and infusion. Augmentation involves dedicating a portion of the math class—typically some prespecified amount of time per week—to life skills math topics that are not specifically indicated but that relate to the content being covered. This way of covering life skills topics is similar to the unit approach. The life skills topics that are chosen are based on their relationship to the content being taught and the importance of the topic to students' lives. Infusion differs from augmentation in that life skills topics are addressed on an opportunistic basis. Based specifically on content covered in the math class, the infusion technique works by relating real-life applications to the content of the textbook, or other materials, being used in the class. Table 10.3 provides both secondary- and elementary-level examples of augmentation and infusion techniques.

Life Skills Mathematics Coursework. Some schools offer specific subject area coursework that is tailored completely to adult needs. In the past, such coursework has been associated with noncredit, alternative programs of

Table 10.3
Augmentation and Infusion Examples

Source	Topic covered	A/I	Sample activities
Practical math textbook (secondary level)	"Budgeting for Recreation"	A	Add coverage on the "economics" of dating
		I	Identify best time and cost for going to a movie
	"Credit Card Math"	A	Add coverage of how to get the best deal on a credit card (e.g., low APR, no annual fee)
		I	Present ways to get a lower APR or waiver of annual fee
	"Maintaining a Vehicle"	A	Add coverage of the realities of being involved in an accident and what one needs to do
		I	Discuss the importance of keeping tires inflated at the proper levels
Basal math textbook (elementary level)	"Using Decimals: Adding & Subtracting Money"	A	Add coverage of costs of purchasing or renting camping gear
	—buying a sleeping bag	I	Discuss where one can buy or rent a sleeping bag
	"Using Tables to Solve Problems"	A	Add coverage on how to use the weather map from the newspaper
		I	Identify other tables that have numbers

Note. A = augmentation; I = infusion.

study, in self-contained, special education settings with students who were not in diploma-track programs. However, variations of this type of coursework for a range of students exist in certain schools where innovative practices have been initiated (see Helmke, Havekost, Patton, & Polloway, 1994).

Life skills coursework in the area of mathematics can be developed along a number of strands. Nearly all secondary schools in the United States have courses entitled "Practical Math" or "Consumer Math," which can be very functional and appropriate for the non–college-bound student. These types of courses can be made more relevant to students if locally derived content is integrated into them. When this is the case, little reason exists for arguing that other math-related life skills coursework is needed.

Other types of math courses or programs can be developed with a clearly functional, life skills orientation. For instance, a comprehensive life skills curriculum, whether it results in a high school diploma or a certificate of attendance, might include coursework with such math classes as "Personal Finance" (where the focus is on personal money transactions and management) or "Survival Math" (a class dedicated to the nonmonetary math needed at home and in the community). Some schools have developed 2- to 3-year math programs for non–college-bound students that focus on the functional math required in daily situations. Table 10.4 describes programs employing both these options.

The fear frequently associated with life skills courses is that they are watered-down versions of more traditional courses. Although this fear may be a reality in some circumstances, it is also possible to develop comprehensive, challenging courses that do differ in orientation from the more traditional math courses but cover essential life skills content. The argument is that all students should take some life skills math courses in addition to more traditional math coursework. Math coursework that is distinctively relevant to students' current and future needs is highly desirable—and a balance between the real-life needs of students and the mathematics standards identified by schools is desperately needed. Having such coursework available as part of the curricular menu makes sense, given what we know about the outcomes and adult needs of students with learning disabilities (Wagner et al., 1993).

FINAL THOUGHTS

The time is right for reexamining how we prepare students with learning disabilities for the demands and challenges of adulthood. Life skills

Table 10.4
Sample Applied Life Skill Math Programs

Program	Purpose	Features
Applied Mathematics Program (Chambers & Kepner, 1985)	Description of an integrated mathematics program for non–college-bound students in the last 3 years of high school.	• integrated curriculum approach • covers goals for arithmetic, algebra, geometry, statistics, probability, and problem solving • emphasizes use of calculators and computers • emphasizes applications to real-world problems
This Is Your Life (Lindsay, 1979)	Students project themselves into the 2 years after leaving high school with no plans for further education but holding down a job and living. Exposes them to true and familiar situations from the adult world; helps students realize the skills they need.	• ongoing "Payday" theme • six units simulating adult responsibilities (i.e., shopping, $1,000 face-lift, "wheels for real," spring vacation, first place, and career bound)
Economics and the Real Life Connection (Murphy & Walsh, 1989)	Description of a program that describes the impact of economics and consumerism on day-to-day life.	• introduction of basic economic concepts • impact of choices we make on economics • decision-making process
Integrating Math Skills Into Vocational Education Curricula (Pickard, 1990)	Description of classroom math instruction in a vocational setting.	• reinforces academic concepts and suggests hands-on activities
Noncareer Mathematics (Sonnabend, 1985)	Description of the most common noncareer mathematics topics that adults need (consumer math, media math, applied math in leisure and theme, and mathematical content and processes).	• outlines several noncareer mathematics units, such as surveys, federal taxes and budget, economics, world population growth, inductive and deductive reasoning, diet, and games
Civics Mathematics (Vatter, 1994)	Description of a program designed to teach math in the context of important social issues of concern to teenagers.	• employs current data • emphasizes newspapers • students are encouraged to pursue projects that may be of service to the community • students gain the understanding that they are part of school, local, and world communities

instruction can and should be included in the curriculum for all students from kindergarten through high school. Students with learning disabilities can greatly benefit from exposure to this type of content. Math instruction should, at the very least,

- Relate to student's current and future needs;
- Be presented in a cohesive fashion;
- Be tailored to a broad range of students, ranging from those who are gifted to those with significant challenges;
- Be integrative by nature—blend well into other content areas;
- Utilize classroom as well as community-based settings—particularly ones in which the individual will probably function;
- Emphasize everyday problem solving.

It makes sense to ensure that students with learning disabilities be competent in applying the math skills needed to survive daily struggles and to be successful—however, success is defined by those individuals.

The message is simple. The math needs of students with learning disabilities must be addressed realistically, functionally, and comprehensively. If this idea becomes the operative guiding principle, then it is extremely likely that a more sensitive and enlightened view of mathematics programming for students with learning disabilities will emerge.

11. Preparing Teachers to Teach Mathematics to Students with Learning Disabilities

RENE S. PARMAR AND JOHN F. CAWLEY

The available developmental data on mathematics performance suggest that students in the general category of learning disabilities (a) function two to four grades below expectancy across mathematics topics (Cawley, Fitzmaurice, Goodstein, Kahn, & Bates, 1979), (b) demonstrate growth patterns in mathematics of 1 year of grade equivalent for every 2 or more years of school (Cawley & Miller, 1989), (c) attain grade-equivalent levels approximating fifth to sixth grade at the time they are leaving school (Cawley, Kahn, & Tedesco, 1989), (d) achieve about 1 year of grade-equivalent growth from the 7th through 12th grades (Warner, Alley, Schumaker, Deshler, & Clark, 1980), (e) demonstrate only limited proficiency on tests of minimum competency toward the end of secondary school (Algozzine, O'Shea, Crews, & Stoddard, 1987; Grise, 1980), and (e) show strange habituated error patterns (Cox, 1975; Lepore, 1979; Pelosi, 1977). A recent analysis of 1,200 KeyMATH scores for students with learning disabilities (Cawley, Parmar, & Smith, 1995) showed an increase of one raw score item per year for each subtest. This leads to an increasing age-in-grade discrepancy over a 6-year period. The KeyMATH data show a trend in which students with learning disabilities (LD), rather than "catching up" or reducing the size of the cumulative discrepancy, are, when contrasted with the test norms, falling farther behind.

We will consider three explanations for the data presented above. It is possible that (a) low performance is intrinsic to the students' structural or functional characteristics and thus little or no change is possible; (b) the instruction presented to students is not designed to meet their unique learning needs and, therefore, difficulties in learning cannot be overcome; and (c) the teacher's curriculum decisions underlying the choice of mathematics are not consistent with the development of understanding by the child. This chapter discusses the issue of preparing teachers to make rapid, effective choices about mathematics (Cawley & Parmar, 1992; Pelosi, 1977) and procedures to effectively teach mathematics to students with learning disabilities, based on the assumption that all students can learn.

The Knowledge and Skills Competencies list compiled by the Division for Learning Disabilities (DLD; Graves, Landers, Lokerson, Luchow, & Horvath, 1993) for teachers of students with LD (see Table 11.1) indicates the need for teachers to be familiar with curriculum, instruction, and assessment in mathematics across the school years. Teacher education programs must evaluate the extent to which they are preparing teachers to meet students' unique needs, so that students with learning disabilities can successfully complete the school curriculum and effectively utilize mathematics in other subject areas and in activities of daily living.

Table 11.1
DLD Knowledge and Skills Competencies in Mathematics

Teachers of students with learning disabilities will demonstrate:

162. Knowledge of K–12 math curriculum
163. Knowledge of a variety of instructional techniques and activities in math
164. Understanding of the appropriateness of various math methods for students who show patterns of error
165. Knowledge of math readiness for math learning
166. Knowledge of the developmental sequence and relationship of the component parts of mathematics as they relate to instruction (e.g., matching of parts to whole and parts to each other, operations, and the decimal system)
167. Understanding the importance of involving the student in setting instructional goals and charting progress
168. Understanding the use of manipulatives and in encouraging students to voice understanding of math concepts
169. Knowledge of the importance of practice, immediate feedback, and review during math instruction
170. Understanding of progressing from concrete to semi-concrete or representational to abstract
171. Understanding of providing a balanced math program (e.g., concept development, computation skills, problem solving)
172. Knowledge of strategies that promote generalization and use of math functionally
173. Knowledge of current research in mathematics instruction

---

Note. DLD = Division for Learning Disabilities of the Council for Exceptional Children.

What do the DLD competencies mean for the preparation of the special education teacher involved in making broad choices about *what* is to be taught, or to the teacher who has to make a multitude of decisions about content and instructional procedures to meet the needs of a diverse group of students? Our perspective is that the answer rests in professional development.

STANDARDS FOR THE PROFESSIONAL DEVELOPMENT OF TEACHERS OF MATHEMATICS

Within the document *Professional Standards for Teaching Mathematics,* published by the National Council of Teachers of Mathematics (NCTM, 1991), is a discussion of teacher preparation. Six standards that highlight the role and responsibilities of individuals engaged in preservice and inservice preparation of teachers are presented. These six standards form the framework of this chapter, with a particular focus on preparation of teachers who will work with students with learning disabilities.

Standard 1: Modeling Good Mathematics Teaching

With reference to Table 11.1, three competencies have particular relevance for this standard. They are *knowledge of the K–12 math curriculum* (No. 162), *knowledge of a variety of instructional techniques and activities in math* (No. 163), and *knowledge of the current research in mathematics instruction* (No. 173). Modeling good mathematics teaching in our methods courses requires developing an understanding of the K–12 curriculum. It is insufficient to focus on one topic or one level of math. In recognition of the time element, it might be advisable to select a topic—for example, statistics—and represent statistics from pre-K through secondary school, or the extent to which students with learning disabilities would likely benefit from study of the topic.

A collaboration is desirable between special education and mathematics education, whereby the techniques and findings of both fields would be shared and interrelated. One component of preservice preparation could involve co-teaching between special education and general education teachers in general education classrooms, resource centers, or self-contained special education classrooms. This would suggest the need for cooperative training programs at the undergraduate and graduate levels. These arrangements would vary with the students' levels (i.e., primary, intermediate, junior high, or senior high) and with the corresponding mathematical knowledge of the teachers. Some models observed include (a) co-teaching, wherein general education and special education teachers share the actual teaching responsibility; (b) assisted teaching, in which the general classroom teacher presents the content and the special education teacher assists all students; and (c) parallel teaching, in which the general education teacher presents the content and the special education teacher takes the students with learning disabilities aside and works with them (Piccillo, 1994). A collaborative relationship might begin with college faculty in special education and mathematics education working together, as is characteristic of some of our own work (from Cawley & Goodman, 1968, to Parmar, Frazita, & Cawley, 1996).

An expected outcome of co-teaching would be enhanced quality of education for all students as they attain higher levels of problem solving, reasoning, communication, connections, and efficiency in their mathematics. The most singular suggestion we could offer to attain this would be for the "methods" courses to redirect themselves to teaching about mathematics rather than teaching how to do mathematics. In the case of a topic such as division, the course might introduce the student to division during the preschool years. Here, the student would group sets of objects by placing so many objects in each group to determine the number of groups (e.g., "Put four cows in each pasture and tell me how many pastures have four cows"), or by placing objects in a specified number of groups to find out how many objects are in each group (e.g., "Put the same number of cows in each of the four pastures and tell me how many cows are in each pasture"). The preschool student would learn about division and intuitively differentiate the principles of *measurement* and *partitioning*. As the student matured, the teacher could teach about problem solving by, for example, asking the student which arrangement is better when packaging a set of toys, three groups with eight toys in each group, or four groups with six toys in each group.

As students approach the time they will learn division, the teacher might present a variety of manipulative representations of division. These could be as follows:

******   1111

with the student manipulating the representation in response to the query "How many groups with two in each group can you make for each pile?" Here the student is dividing a three-digit representation by a single-digit representation. The teacher could then label each symbol (e.g., # = 100, * = 10,

1 = 1) and write this with the students as 400 + 60 + 4. The teacher could ask the students for another way to write the number and lead them to:

100, 100	10, 10	1 1 1 1
100, 100	10, 10	
	10, 10	

then ask, "How many groups with two in each group can you make in each column?" The students could count and show two sets of 100 with two in each set, three sets of 10 with two in each set, and two sets of 1 with two in each set. They have divided 464 by 2 and responded with 232. Symbolically, the teacher writes

$$2 \overline{)4\ 6\ 4}$$

and directs the students to the same question: "How many sets with this many [points to 2] can you make in each column?"

Where the aforementioned becomes more meaningful and contro-versial is in an item such as $3 \overline{)1\ 5\ 6}$, for which most teachers inform the students that "3 into 1 does not go." However, we realize this is not so if the item is represented as $3 \overline{)100 + 50 + 6}$, where we clearly see the place value representation of the 1 as 100—and we all know that 3 will go into 100 about 33 times. We need our teachers to understand this and be able to explain why the statement "3 into 1 does not go" is misguided, even if it is fast to do. The better question would be, "How many sets with this many [point to the 3] can you make here [point to the 1]? The student will tell you "none." With the appropriate background and understand-ing of place value, the student will respond, "I will rename the 100s into 10 sets of 10s and group them with the 5 to make fifteen 10s." The stu-dent can then look at the 10s column, where there are now 15 tens, and respond to the question, "How may sets with this many [point to the 3] can I make here [point to the 15]?" with "five." Both the teacher and the student now understand why we "move over," and both can see that divi-sion can be done without any multiplication or subtraction at a much higher level of meaning than "3 into 1 does not go."

Not all the competencies represented by Graves et al. (1993) are clear. For example, *understanding the progression from concrete to semiconcrete or rep-resentational to abstract* is of undetermined value, though we recognize some efforts in this regard (Miller & Mercer, 1993), because the paradigm needed for a more effective determination of it (e.g., a Latin-square where all com-binations are presented to all students in all sequences) has yet to be tried.

From the mathematics perspective, often the "concrete" may represent the more astute outcome and, in reality, the higher order outcome.

Individuals engaged in teacher preparation need to consider the extent to which their own instructional practices model effective teaching. If prospective teachers are to be encouraged to use activities and discourse in their classrooms, they must have the opportunity to engage in mathematics learning through activities and discourse. This might include a problem-posing activity, such as, "Why are manhole covers round?" The activity could include brainstorming in small groups and the development of models of manhole covers that are round and square. The student teachers could then determine what happens to square and round manholes when they are tipped on their sides. They could discover that round manholes do not have a diagonal (no hypotenuse) and cannot fall through, which is not the case with the square cover.

Alternatively, if the prospective teachers are to be encouraged to use a modeling approach, they could create models of manholes having a variety of sizes and shapes. They could then show the students what happens to manholes of different shapes and guide the students to a set of principles, such as that round manhole covers will not "fall through" whereas square ones will. The teacher could then demonstrate various relationships between circular and square regions of the same side length (diameter) and the relative areas of each. The teacher would then engage the students in a culminating activity, such as a generalization task.

The experiences teachers have during preservice convey strong messages about what constitutes appropriate teaching and learning (NCTM, 1991). The "explain–practice–memorize" paradigm often advocated in methods courses does not provide an appropriate model for teachers and may be a major source of mathematics anxiety and avoidance among teachers (Greenwood, 1984). Our position would be that the teachers should (a) be presented with a variety of instruction models, (b) compare them, and (c) be able to illustrate their strengths and weaknesses.

Standard 2: Knowledge of Mathematics

O'Neill (1988) cited two comments from special educators relative to the matter of curricula. One individual indicated that the curricula for general education and special education are largely the same, at the elementary level, except that special educators try to "break down instruction more finely" (p. 2) than general educators. Another special educator stated, "Just use good, straightforward, direct instruction, in serving mildly handicapped students. We don't need fancy special education curricula" (p. 2). Clearly, neither of these is acceptable. First, math textbooks published after the 1989 reform are on average about 16% (approximately 70 pages) longer than those published before reform (Chandler & Brosnan, 1994). With textbooks running nearly 600 pages in length, breaking them down more finely is going to increase the number of tasks per topic, slow students' progress, and decrease the time for content coverage. Second, the DLD Competencies stress the importance of knowledge of the K–12 mathematics curriculum (No. 162), knowledge of the developmental sequence and relationships among the component parts of mathematics (No. 166), and providing a balanced math program (No. 170), so that teachers will be able to make appropriate curriculum decisions rather than simply following arbitrarily prescribed sequences.

Knowledge of mathematics is more than simply "being good at mathematics." Knowledge of mathematics includes (a) understanding the meanings, principles, and processes of a wide range of mathematics appropriate to the needs of the students; (b) recognizing unusual performance on the part of a student and how to adapt activities to determine the basis for this performance; and (c) knowing the developmental characteristics of the student in such detail that individualized curriculum choices can be made as to *when* it is an appropriate time to present certain mathematics to a student, the *sequence* in which it should be presented, the *intensity* or length of time one will stay with a topic to assure mastery, the *mixture* of mathematics that should be presented, and *how* to determine that a student has attained proficiency and mastery of the principles.

Teacher education programs are often woefully inadequate in presenting adequate coverage of curriculum and instructional techniques in mathematics. In a survey of the syllabi of more than 250 methods courses in special education preservice programs, Lucas-Fusco (1993) found that although, on average, 6.36 class sessions per semester in methods courses for learning disabilities were devoted to methods of teaching reading, only 0.57 sessions were devoted to mathematics instructional methods. While recognizing the importance of reading and literacy skills, we must also be aware that a number of studies of minimum competency have shown that students with learning disabilities perform much better in reading than in math. Yet, they are expected to pass both components of the tests to get a high school diploma (Parmar, Klenk, & Cawley, 1996).

One technique to provide adequate coverage of a variety of areas in methods courses, given the limited time, is the integration of various skill areas. Strategies for developing literacy can be implemented within the context of mathematics and other content areas. Thus, students with disabilities can read story problems, write their own problems, and discuss and write justifications and explanations for their mathematical solutions. Science and mathematics also have natural connections through classification, measurement, and data recording. Such an approach is consistent with the NCTM (1989) process standards of communication and connectedness in mathematics.

To make appropriate decisions regarding content for students with learning disabilities, teachers must become familiar with state curriculum guidelines and frameworks, especially as these frameworks encourage more open-ended forms of mathematics. They must also consider prioritizing topics so that more important ones are covered thoroughly and less important topics are given less time. For example, students are likely to utilize data and graph interpretation in adult life but are less likely to actually compute three- and four-digit multiplication and division without the use of a calculator. It is more beneficial for teachers to spend time on the former than the latter. Preservice and inservice programs must include discussion of curriculum priorities, based on the variety of adult careers and professions.

Standard 3: Knowing Students as Learners of Mathematics

Fennema and Franke (1992) highlighted the importance of understanding the learner's cognitions in order to design effective instruction in mathematics. DLD Competencies also emphasize the need for teachers to be familiar with assessment in general, as well as be able to modify instruction for student errors in mathematics specifically (No. 164). We will illustrate three possible clusters of students with disabilities in mathematics. (We recognize that there are many other types of difficulties presented by the students; these three are for purposes of illustration only.) Our assumption is that meeting the needs of each student enables us to meet the needs of all students. In keeping with the basic tenets of *specially designed instruction* as defined in P.L. 94-142, we use these illustrations to call attention to the variability and uniquenesses among students.

The three clusters used for illustration are, first, a group that manifests overall delays in all aspects of achievement (e.g., Warner et al., 1980). These students are discrepant in reading and language as well as mathematics and therefore show a consistently below-average achievement. Second is a group that is discrepant in mathematics specifically, in the sense that they do not master concepts or routines, nor do they remember their "facts" (Pieper & Deshler, 1980; Rourke & Strang, 1983; Webster, 1980). A third group is characterized by the habituation of unusual and incorrect algorithms and procedures (Cox, 1975; Lepore, 1979; Pelosi, 1977).

Students with Overall Delays in Achievement. For students demonstrating low achievement across various academic areas, performance in mathematics is often interrelated with difficulties with reading, poor attention, difficulty with memory, or difficulty integrating new information with existing knowledge structures.

We offer four guidelines to address the needs of students who demonstrate overall delays. First, help teachers to clearly identify the important mathematics and conceptualize this mathematics into a curriculum that covers 12 or more years, so that the time in school can be used efficiently to acquire necessary capabilities and knowledge in mathematics. For example, if our curriculum is to focus on "big ideas" (Carnine, 1991)—a practice we heartily support—then obviously it is essential to identify those big ideas. Using illustrations for the topic Volume presented by Carnine (1991), big ideas would be, for example, "Why do you square the radius in πr^2 when determining the area of a circle?"; and "If the area of a circle is always less than the area of a square, what is the relationship between the volume of a rectangular prism and that of a cylinder if the two figures have base areas and heights that are the same?"

Second, we need to help teachers shift the emphasis from teaching predominantly via paper-and-pencil worksheet (symbolic format) activities to alternative representations, such as pictures (visual format) or manipulatives (activity format). These may be more important relative to students' responding to present forms of assessment (Lane, 1993; National Assessment of Educational Progress, 1992) than for their initial learning. For those students who can demonstrate a clear understanding of an operation in a problem-solving or application activity, the hand-held calculator ought to be the primary means of computation. Third, emphasize the development of meanings and understandings through problem-solving activities that recognize the importance of statistics, geometry, and measurement in our activities of daily living; develop the skills needed for these topics with practice that is applied to realistic situations. Finally, conduct activities that enable students to see the connections between school activities and what they learn in other areas of academics as well as the real world. For example, students could apply statistics (e.g., frequency distributions) to construct graphs to show the various voting patterns of members of the town council, sports statistics from the newspapers, or the sale of car models at a local dealership.

Students Discrepant in Mathematics Specifically. For students who demonstrate discrepancies in mathematics only, the focus of instruction should be identification of their areas of difficulty through individualized assessment, and the implementation of instruction that addresses the specific difficulty. We suggest that teachers be taught to work within the framework of the principles of *least error correction* and *proximity analysis*. Least error correction stipulates that one will identify and correct only the action that is the cause of a specific difficulty. If more than one action is involved, they will be corrected in sequence. For a simple illustration, examine the following subtraction item:

$$
\begin{array}{r}
473 \\
-134 \\
\hline
300 \\
20 \\
\underline{9} \\
319
\end{array}
$$

Note that the student used a left-to-right algorithm and used it correctly. He or she made a careless error in subtraction in the 10s column, which should be the focus for remediation, not the algorithm, as the algorithm was correctly used. Yet, in many cases, the student is told to change the algorithm (e.g., start with the right column) as an intervention.

Proximity analysis involves an examination of the error in relationship to previous items that are correct. Table 11.2 presents items extracted from a content sequence of 41 subtraction items. The illustration in the table shows that the student was correct in items S6 through S11. An examination of S12 shows that renaming in the 10s column resulted in a zero. The student applied an incorrect algorithm to the 10s column in that he or she subtracted the larger number from the smaller number instead of renaming from the 100s column. This provides a very specific error to correct.

The same principle can be applied across all topics. For example, effective curricula should include the three basic ideas of subtraction as (a) What must be added to a number to make it as large as another number? (b) How much larger is one number than another? and (c) What remains of a number after part of it has been removed? Note that the first two ideas present subtraction as a search for the missing addend and, in a sense, present subtraction within a framework of addition (e.g., $2 + \_ = 6$). The approach that presents subtraction as only $6 - 4 = 2$ is likely to limit the student's long-term problem-solving experience, especially when presenting problems of the "start unknown" type (Riley, Greeno, & Heller, 1983, p. 182) or the indirect type (Parmar, Cawley, & Frazita, 1996), such as, "A boy has 3 apples *left* after he *gave* 2 apples to a girl; how many apples did the boy start with?"

Table 11.2
Content Sequence for Subtraction

Student performance			Descriptor
S6	565 − 2 563	867 − 5 862	Three-digit minus single-digit; no renaming
S7	857 −24 823	975 −43 932	Three-digit minus two-digit; no renaming
S8	471 − 4 467	472 − 3 489	Three-digit minus single-digit; renaming 1s
S9	763 −38 725	872 −44 828	Three-digit minus two-digit; renaming 1s
S10	345 −63 282	828 −54 774	Three-digit minus two-digit; renaming 10s
S11	632 −37 595	826 −58 768	Three-digit minus two-digit; renaming 1s/10s
S12	612 −28 624	613 −47 646	Three-digit minus single-digit; renaming 1s/10s// zero 10s

Students Who Have Habituated Incorrect Procedures. These students tend to achieve poorly in mathematics as the result of poor conceptual understanding, due to the habituated use of incorrect algorithms (Cox, 1975) or the use of strange and illogical algorithms (Lepore, 1979). One excellent example of the former, and, unfortunately, a difficult one to correct because it often works, can be found in Pelosi's (1977) work. Pelosi had given a student an item similar to the following:

$$\overset{\displaystyle 7}{\underset{}{\overset{}{6\overline{)4\,2\,6}}}}\,\overset{1}{}$$

The student divided 6 into 6 and obtained 1; the student then divided 6 into 42 and got 7, an error that is likely due to having been taught to "start on the right." Not knowing whether the student was simply using an alternative algorithm or was confused, Pelosi immediately gave the student another item, similar to the one below, and got the following type of response:

$$
\begin{array}{r}
6 \quad r1 \\
1 \\
\hline
7\overline{)437}
\end{array}
$$

The student said, "Seven into 7 is 1. Seven into 43 is 6 and 1 left over." Here we clearly see that the student was using the algorithm incorrectly, though it initially appeared that he was applying the correct procedure.

Cox (1975) showed us that students with learning disabilities who incorrectly use an algorithm habituate the use of that algorithm and repeat the mistakes. Teachers need to know how to intervene before a procedure is habituated. If a student consistently completes 30% of items correctly in a given set of problems, then he or she is practicing incorrect approaches 70% of the time. Lepore (1979) highlighted another concern, regarding times when students use strange and bizarre algorithms or procedures and often describe them as though they were acceptable and may even argue for them in the face of controverting evidence. Teachers must be educated so they can respond to students' illogical explanations and arguments and adapt instructional activities to challenge misconceptions. One approach may be to ask the student to solve a number of similar problems using manipulatives and other representations. Many incorrect algorithms persist in paper-and-pencil applications but will not work with manipulatives.

Standard 4: Knowing Mathematics Pedagogy

If the instructional goal for students with learning disabilities is for the students to memorize the multiplication tables, or to do subtraction only one way, then we can continue our present instructional practices. If, on the other hand, students with learning disabilities are viewed as potential problem solvers, persons who can develop meaningful concepts and principles, and individuals who can be active rather than passive learners, then changes are needed in instructional practice.

The great majority of the mathematics intervention research of special education does not directly address changes in the nature of mathematics or instruction in mathematics. Rather, the changes have been in the management and motivational features of instruction (e.g., goal setting, modeling, drill and practice, on-task behavior, time-delay methods) that have a common focus on the development of the skills and routines of arithmetic computation. To illustrate, consider research on one dependent variable (arithmetic computation) and an independent variable such as time delay (Koscinski & Hoy, 1993). The instructional procedure is to present the students with flashcards and have the students memorize the number combinations. The students are assigned to one or more treatments or a control group. The mathematics instruction is the same; it is the condition under which the time-delay treatments differ that constitutes the intervention.

Support for our observations is found in an article by Mastropieri, Scruggs, and Shiah (1991), who concluded that although the mathematics interventions they reviewed reported successful outcomes, few related the intervention under study to any specific theory or characterization of learning disabilities, or to mathematics. Most studies were based on a behavioral orientation in which the dependent measure was generally number or percentage of items or digits correct, often within a set time limit. No studies were reported that used any alternative perspectives, such as developmental or constructivist models. In addition to some of the shortcomings observed by Mastropieri et al., they further observed that the studies (a) were largely short-term interventions; (b) focused almost exclusively on arithmetic computation or arithmetic word problems; and (c) frequently used scores on arithmetic tests only as an external measure when studying a particular behavioral intervention, rather than studying the learning of arithmetic per se (Parmar et al., 1996). Among these are reports of teaching single-digit arithmetic computation to students who are 16 years of age.

The following question arises: If we continue to stress the teaching of computation skills to 16-year-old students, when will they have opportunities to engage in higher levels of problem solving? And, if the intervention research continues to focus on arithmetic computation, who is to validate treatment effects for geometry, measurement, and statistics, all of which seem more important in terms of real-world applications than arithmetic computation (Parmar & Cawley, 1995). What might be the outcome if a group of students learned multiplication from the perspective of arrays and learned that there are only three tables to remember (i.e., 2s, 5s, and 10s) and that all others are permutations of these (see the case of "Charles" in Cawley & Parmar, 1992)? Teacher educators and teachers need to go outside the special education literature to find illustrations and guidelines for programming in mathematics. Two suggestions would be to encourage teachers to join professional organizations, such as the National Council of Teachers of

Mathematics, and examine the literature on state and national standards for mathematics.

Standard 5: Developing As a Teacher of Mathematics

For changes to occur in how mathematics instruction is viewed, it is important to examine the theoretical frameworks that have led to current practices. As we look back on the development of instructional theories of mathematics instruction, we can reasonably group them into three broad classifications:

1. Behaviorism (e.g., Skinner, 1968; Thorndike, 1906);
2. Guided Meaning (e.g., Bruner, 1966); and
3. Constructivism (e.g., Montessori, 1964; Piaget, 1965).

The behaviorists' general interpretation of instruction is that activities should be created that solidify the bonds between stimuli and responses (Thorndike's view as described by Resnick & Ford, 1981, p. 15). For example, the student learns to say "six" when presented with 2×3. From this perspective, mathematics is perceived as having specific structures, and drill-and-practice activities are highlighted. Automatization and time-per-task outcomes are important. Memory is an integral element of behaviorism. The primary instructional material is a form of worksheet or flashcard that is used first for acquisition and next for rehearsal and practice. This has typically been the format of the majority of software used in microcomputers (Cosden, 1988). The teacher assumes a dominant role so as to accept responsibility for student progress and to be able to make instructional changes when indicated. Generalization is developed through demonstration of the skill under varied circumstances.

Guided-meaning enthusiasts seek relationships between and among combinations of meanings and outcomes. The statement 3×2 is perceived as a relationship between two concepts that form a principle; 3×2 evolves as a *factor-by-factor* = *product* relationship and multiplication and division are taught simultaneously. The creation of activities that require the student to explain, prove, and justify task outcomes is a common theme in this approach. For example, for younger students, this might involve the relationships between students and apples (e.g., the number of apples a student will receive, or the number of students who will receive a fixed number of apples); for older students this might involve activities for rate and distance (d = rt), whereby students might determine *rate of travel, distance,* or *time.* The teacher is an integral element of the activity and (a) provides guided practice and extends generalizations; (b) prevents habituated error; (c) asks questions to highlight meanings; and (d) dialogues with students to bring about proof, reasoning, and elaboration.

The constructivists focus their activities so as to engage the student in ways that lead the student to construct meanings (Cobb, 1994; Kamii & Lewis, 1993). Students are free to interact with a variety of materials, to ask questions, and to offer alternative solutions. The constructivists generally push for deeper understandings and meanings and initially are less concerned with speed of response and accuracy. The primary instructional materials are manipulatives and real-life settings. Generally, algorithms are not taught directly; rather, the student is provided with many opportunities to invent or discover them. The teacher is more of a facilitator than a leader. The environment provides for authentic and situated learning, and learners are perceived to make up a community in which students help each other. Alternative explanations and views are encouraged.

Aspects of the three theoretical perspectives of mathematics instruction in special education are shown in Table 11.3. It is likely that there are advocates for each and that these advocates will continue with their preferences, independent of any data to the contrary. The appropriateness of different approaches may vary with content. All teachers of special education should be presented with a fair and unbiased representation of a variety of instructional theories, for it is likely that there will be a time when all or parts of each can be used to meet specific student needs.

As teachers develop in mathematics teaching, a number of changes are likely to occur, both in themselves and in their students. It may be that a change in attitude toward mathematics is the most important outcome of any program. Any program that addresses math anxiety among teachers is likely to be a program that increases both teachers' knowledge and their instructional capabilities. In fact, preservice teachers who are highly anxious about mathematics tend to be more understanding of student difficulties, more likely to slow the pace of instruction, and more willing to review and reteach topics than are teachers who are not highly anxious (Unglaub, 1995). The very fact that teachers can observe college faculty in special education enjoying math, displaying a wide variety of knowledge of mathematics, and employing interesting and challenging procedures to teach mathematics is fundamental to their own success.

Standard 6: The Teacher's Role in Professional Development

There are three primary roles for the teacher in professional development. One role requires active participation in workshops and training ses-

Table 11.3
A Comparative Chart of Mathematics Instruction Frameworks in Special Education

Behavioristic	Guided meaning	Constructivist
A focus on developing skills	A focus on deep levels of understanding	A focus on deep levels of understanding
Identification of a single algorithm or procedure to be taught	Presentation of a variety of materials and activities	Evolvement of a variety of materials and activities
Teacher provides a sequence of steps (task analysis) to solution	Teacher actively encourages discussion about topic under study	Teacher-facilitator encourages discussion about topic under study
Students memorize and follow set procedures to solution	Students interact with materials to develop conceptual knowledge	Students interact with materials to develop conceptual knowledge
Mastery of skills prior to their application in problems	Teacher provides models of problem-solving strategies and processes	Students learn through discovery of mathematical principles and concepts
Individualized drill and rehearsal activities for mastery	Students apply strategies and processes to solve novel problems	Students assimiliate new concepts into prior knowledge
	Students are provided with moderately novel problems to reinforce learned concepts	Students change mental structures to accommodate new knowledge

sions and the reading of professional literature. A second role involves translating both student and teacher needs into professional development activities at the school, district, and state levels. A third role is developer and leader of professional development activities for other teachers and staff.

We are presently involved in a 5-year effort (Parmar, Cawley, & Frazita, 1993) to enhance the assessment capabilities of general and special education teachers on behalf of students with disabilities, and to use alternative assessment as a way of enhancing instructional practice. The project includes five two-member teams of mentor teachers, each team consisting of a general education teacher and a special education teacher. These teams participated in a 1-week summer inservice program, followed by a series of 10 sessions during the school year, in addition to classroom implementation and follow-up. Through the inservice, the teachers acquired knowledge about varying approaches to assessment and how those approaches should interact with or guide their instruction. All teachers were presented with the NCTM (1993) *Assessment Standards* and a variety of supporting literature and assisted in developing alternative assessment models appropriate for their teaching situations. The mentor teachers then invited two colleagues from their schools to participate in the following year's training. The outcomes of the first year were presented at the state CEC conference by the mentor teachers (Senio et al., 1995). Each teacher provided an illustration of procedures used in their classrooms and data from assessment that consequently influenced their instruction. The long-term implementation of a mentor-teacher model appears to have some benefits in producing substantive change in teacher beliefs and competencies.

SUMMARY

The field of learning disabilities needs to give greater consideration to preparing teachers to work with students individually and in included settings. Courses in curriculum and methods for teaching students with learning disabilities must provide good models for teachers to use in their classrooms. Activities need to be designed around real-life problem solving and "big ideas" in mathematics. Preservice teachers need to be assisted in their efforts to develop as professionals through sustained collaborative work with faculty mentors, focusing on classroom issues of relevance.

Further, considerable collaborative effort is needed to develop solutions to the differences in curricular, instructional, and assessment perspectives of general education mathematics specialists and special educators. We suggest two possibilities. First, one or more of the major organizations could organize/co-sponsor a small working conference focusing on *mathematics for all students.* Such a conference could be organized by DLD, the Council for Learning Disabilities, CEC, or NCTM. Second, we suggest that the staff of each college or university having both specialities get together within their units and restructure the courses that prepare teachers to teach mathematics. The focus should be the enhancement of mathematics for *all* students.

NOTE

Preparation of this chapter was made possible, in part, by a grant from the U.S. Department of Education, Office of Special Education and Rehabilitative Services. The views expressed are those of the authors.

12. Achieving Meaningful Mathematics Literacy for Students with Learning Disabilities

SUSAN R. GOLDMAN, TED S. HASSELBRING, AND

THE COGNITION AND TECHNOLOGY GROUP AT VANDERBILT

This chapter completes this book on mathematics and learning disabilities. Unlike the other authors, who were asked to assess the state of the field in specific areas, we were asked to provide some comments on the future. While envisioning the future, we were asked to consider the role(s) that technology might play. In crafting our vision of the future, we made the assumption that a primary goal of mathematics instruction for youngsters with learning disabilities is the achievement of basic mathematical literacy. Of necessity, then, the starting point for envisioning the future is to revisit an issue addressed by a number of the chapters: changing standards for mathematics learning, and thereby changing definitions of basic mathematical literacy.

The topic of changing standards in mathematics quite naturally leads into a discussion of theories of cognition and learning that provide the psychological and educational context for thinking about basic mathematical literacy. We argue that the shift from behaviorist learning theories to constructivist and social constructivist theories (see Rivera, chapter 1 of this book) provides an opportunity to forge a hybrid view of mathematics instruction for the learning disabilities field. This hybrid embeds, or situates, important skill learning in meaningful contexts (Bransford, Goldman & Hasselbring, 1995; Cognition and Technology Group at Vanderbilt (CTGV;

1996a). We discuss some examples of instructional approaches to complex mathematical problem solving that make use of meaningful contexts and various ways in which technology is important to these contexts and to realizing hybrid instructional models.

EVOLVING CONCEPTIONS OF MATHEMATICAL LITERACY

One of the important issues highlighted by this book is the impact of changing conceptions of what it means to be mathematically literate. This change has been stimulated by and reflected in such policy-level events as the National Council of Teachers of Mathematics' (NCTM) Standards (NCTM, 1989), as well as Goals 2000 and the New Standards (1995) project. For example, the NCTM Standards were developed in the context of glaring evidence of the need for major improvements in our nation's educational system (e.g., Resnick, 1987). Especially important has been the emphasis on the need for *all* students, not simply a select few, to learn to solve problems, reason, and learn on their own (e.g., Bransford, Goldman, & Vye 1991; A. Brown & Campione, 1990; Nickerson, 1988; Resnick, 1987; Resnick & Klopfer, 1989; Scardamalia & Bereiter, 1991). This mandate is an inclusionary one, and its implications generate strong reactions from special educators. We address two issues with respect to the standards: Do we need them? (we think we do), and, What precisely do they have to say about instruction?

The Need for New Standards

Changes in conceptions of mathematical literacy are occurring in the midst of societal recognition that 20th-century skills will not be sufficient for the workforce of the next century. The Secretary's Commission on Achieving Necessary Skills (SCANS) report issued by the U.S. Department of Labor (1992) clearly identified collaborative activity, problem solving, communication, self-assessment, and competence and confidence with technology as critical to success in the 21st century. The emphasis on *all* students needing to be able to solve problems, reason, and take charge of their learning should be cause for celebration, because students with learning disabilities are not being excluded from opportunities to develop the life skills needed for success in the next century.

Often overlooked in discussions of the SCANS report is that basic literacy in traditional academic content areas, including mathematics, is assumed. A point that seems to elicit concern among special educators is that the SCANS report, as well as the new Standards, broaden the definition of basic mathematical literacy to include a wider range of competencies than has traditionally been the case. This is as it should be, because *all* students need to be mathematically literate to the point that they can "figure out" math-related problems they encounter at home or in future work situations. These rarely look like the problems presented in most mathematics texts and workbooks.

Unfortunately, there is considerable evidence to indicate that even the traditional standard of basic mathematical literacy is not being attained by many of our students, whether they be general education students or students exhibiting learning problems. Since the beginning of the National Assessment of Educational Progress (NAEP) two decades ago, media, parents, government officials, and educators have debated potential causes for the low test scores in America's schools, and have searched for ways to remedy the problem. Although American students do well on whole-number computations, they have difficulties with fractions, decimals, and percentages and with problems that involve unfamiliar, nonroutine tasks. Word problems that involve two or more steps are particularly problematic for these students (Kouba et al., 1988). And, although American youth possess a fairly good knowledge of the *procedures* associated with rational numbers, probability, measurement, and data interpretation, they lack the conceptual knowledge that enables them to *apply* their knowledge in problem-solving situations (C. Brown et al., 1988a, 1988b).

For students with learning disabilities, the situation is even more alarming because much of their instruction focuses largely on procedural computation skills that are executed in similar ways each time they are used. However, studies show that students with learning disabilities experience greater difficulty than their same-aged peers with both compution and word problems (e.g., Ackerman, Anhalt, & Dykman, 1986; Cawley & Miller, 1989; Fleischner, Garnett, & Shepherd, 1982; Goldman, Pellegrino, & Mertz, 1988; Russell & Ginsburg, 1984). This situation exists despite the fact that the bulk of instructional time is spent learning procedures for doing mathematics computation and simple word problems. (For further discussion, see Babbitt & Miller, 1996, and Carnine, Jones, & Dixon, 1994.)

Students' deficits in traditionally defined mathematics skills, in combination with the growing awareness of the insufficiency of these skills in preparing individuals for the 21st century, create a climate for evolving conceptions of basic mathematical literacy. But specifying new standards is not the same as specifying curricular content and instructional environments.

Standards Are Not Curricula

A number of special educators have expressed severe reservations about the adoption of the NCTM Standards or the New Standards because they have not been demonstrated to be relevant and valid for students with diverse needs (e.g., Hofmeister, 1993; Jones, Wilson, & Bhojwani, chapter 8 of this book; Mercer, Harris, & Miller, 1993; Miller & Mercer, chapter 4 of this book; Rivera, 1993). Although this concern is understandable, it reflects a fundamental misconception about the Standards. They specify topics, but they are not curricula. They exemplify possible tasks and instructional techniques but fall far short of defining or specifying classroom practices or an instructional program that will lead to attainment of the Standards. In short, the Standards set goals but not the means to achieve them.

The Standards are the conceptual basis for developing instructional programs. They call for a more generative approach to mathematics learning. The NTCM (1989) states,

> (The) mathematics curriculum should engage students in some problems that demand extended effort to solve. Some might be group projects that require students to use available technology and to engage in cooperative problem solving and discussion. For grades 5–8 an important criterion of problems is that they be interesting to students. (p. 75)

The NCTM's suggestions for changes in classroom instructional activities include more emphasis on complex, open-ended problem solving, communication, and reasoning; more connections from mathematics to other subjects and to the world outside the classroom; and greater use of calculators and powerful computer-based tools, such as spreadsheets and graphing programs for exploring relationships (as opposed to having students spend an inordinate amount of time calculating by hand).

The content and performance standards for middle school mathematics being developed by the New Standards group (1995) are similar. They provide standards in several broad concept areas, such as number and operations, functions and algebra, statistics and probability, mathematical communication, and mathematical skills and tools. Under each concept area, more specific competencies are provided. For example, in the statistics area, the standards indicate that students should collect, organize, and analyze data. Under skills and tools, the standards indicate that students should compute accurately with arithmetic operations on rational numbers. As with the NCTM standards, the New Standards are not curricula and do not specify the instructional program to be used to achieve these standards. It is interesting to note that even when curriculum materials are specifically designed to support the kinds of activities described in the NCTM guidelines and in the New Standards, they can be used in traditional "drill-and-kill" ways (CTGV, 1992b). Thus, in addition to important instructional program development, there is important work to be done on implementation of such programs.

In thinking about program development and implementation, it is important to point out that the standards do not mean that educators should single-handedly "throw out" all current methods of mathematics education. Nor do they mean that educators should not teach and that students should be left to "discover" mathematics through hands-on activities.

Rather, the fundamental implications of the standards are that children need to actively participate in learning environments and make constructive, generative contributions to what they learn and to assessments of it. This type of participation in learning implies that students need to develop the skills that will allow them to direct their own learning more adequately. This is true for all students, including students with learning disabilities.

Active involvement in learning is less likely to happen, however, if school math is taught in a decontextualized manner, with little emphasis on learning how math can be used or why it is important (CTGV, 1990, 1993b, 1994a, 1994b, 1996a, in press-a). This form of school math is very different from the informal mathematical knowledge that students develop prior to entering school. Increasingly teachers, curriculum developers, and researchers are recognizing that children have considerable knowledge related to mathematical problem solving when they enter school (Peterson, Fennema, & Carpenter, 1991; Resnick, Bill, Lesgold, & Leer, 1991; Vye, Sharp, McCabe, & Bransford, 1991). Students come to school with at least two kinds of intuitive knowledge related to mathematics learning: (a) a knowledge about amounts of physical material and the relations among these amounts (e.g., some materials can be cut into pieces and then "added" back together again) and (b) rules for counting sets of objects (Peterson et al., 1991; Resnick et al., 1991). As Ginsburg (chapter 2 of this book) contends, this informal knowledge should serve as the basis for developing formal mathematical knowledge.

COGNITIVE PERSPECTIVES ON FORMAL MATHEMATICAL KNOWLEDGE AND LEARNING

Research in cognitive science points to the distinctions among three basic types of mathematical knowledge: declarative, procedural, and conceptual. Each is critical to developing mathematical literacy consistent with the NCTM Standards and New Standards.

Declarative knowledge is best represented as facts about mathematics. It can be conceptualized as a network of relationships containing basic problems and their answers, such as $4 + 7 = 11$ or $11 - 4 = 7$. The facts stored in this network have different strengths that determine how long it takes to retrieve an answer. The stronger the relationship, the more rapid and effortless the retrieval process. For example, if the fact $2 + 3 = 5$ has greater associative strength than the fact $7 + 5 = 12$, it will take less time to retrieve the answer 5 to the first of these two problems (Goldman & Pellegrino, 1987; Pellegrino & Goldman, 1987). Ideally, facts stored in the network are retrieved from memory quickly, effortlessly, and without error. We have known for some time that computer-assisted instruction is effective for increasing fluency of fact retrieval among students with learning disabilities (e.g., Goldman & Pellegrino, 1986; Goldman et al., 1988; Hasselbring, Goin, & Bransford, 1988). Fluency is important because declarative knowledge serves as the building blocks for the second knowledge type, *procedural knowledge.* Effortful retrieval of basic facts or effortful calculation of them prevents attentional and processing resources from being focused on other aspects of mathematical knowledge and problem solving.

Procedural knowledge can be defined as the rules, algorithms, or procedures used to solve mathematical tasks. Procedural knowledge is represented as step-by-step instructions in how to complete tasks, and the steps are to be executed in a predetermined linear sequence. Thus, in using procedural knowledge, declarative knowledge is accessed to perform procedures for acting on or dealing with a problem situation. Procedural knowledge is composed of two parts: One refers to the symbolic representation of mathematics, such as whole-number operation symbols $(+, -, \times)$; the other consists of rules for completing mathematical tasks, such as algorithms. Procedural knowledge is therefore concerned primarily with an awareness, or surface knowledge, of how mathematical computations are to be performed (Anderson, 1982, 1987).

In contrast to declarative and procedural knowledge, the third knowledge type, *conceptual knowledge,* is defined as a connected web of information in which the linking relationships are as important as the pieces of discrete information that are linked. Conceptual knowledge determines *understanding,* rather than mere working of computational steps (Hiebert & Lefevre, 1986). The development of conceptual knowledge is achieved by constructing relationships between pieces of information. This linking process can occur between two pieces of information that already have been stored in memory or between an existing piece of knowledge and one that is newly learned. The literature of psychology and education is filled with accounts of insights gained when previously unrelated items are suddenly seen as related in some way. Such insights are the bases of discovery learning (Bruner, 1961). We characterize this as an increase in conceptual knowledge.

An illuminating account of this kind of growth in conceptual knowledge in elementary mathematics is found in Ginsburg (1977). Ginsburg described many points in the learning of numbers and arithmetic where understanding involved building relationships between existing bits of knowledge. For example, Jane (age 9) understood multidigit subtraction for the first time when she recognized the connection between the algorithm she had memorized and her knowledge of the positional value of each digit. Relationships can tie together small pieces of information or larger pieces that are themselves networks of sorts. When previously independent networks are related, there is a dramatic and significant cognitive reorganization (Lawler, 1981).

Mathematical literacy requires the development of interactive relationships among declarative, procedural, and conceptual knowledge. The development of these relationships is critical for accessing and using the knowledge to solve problems. If students understand the underlying reasons about "how" and "when" to use a procedure, they will be able to store it as a part of their knowledge network, thus developing links with other pieces of information. Links between conceptual knowledge and procedural knowledge can help students select an appropriate procedure because they will understand the rationale for applying the procedure and will be able to identify situations in which it is appropriate to use the procedure.

Declarative knowledge that is not tied to procedural knowledge, and procedural knowledge that is not tied to conceptual knowledge, are, at best, limited in their usefulness and, at worst, useless for the student. If students do not understand why they are using specific procedures in particular contexts, then there can be little possibility of transfer to other contexts. Relationships must be established among declarative, procedural, and conceptual knowledge to make this knowledge a part of the student's repertoire of skills.

There are several telling examples of the failure to accomplish these interconnections. For example, we conducted work on word problems with fifth- and sixth-grade students who were having difficulty in school, especially in areas of reading and mathematics (CTGV, in press-a). We presented students with written versions of simple word problems, such as the following:

1. Tony rides the bus to camp every summer. There are 8 other children who ride with him. The bus travels 9 miles an hour. It takes 4 hours to get there. How far away is the camp?

2. John is standing in front of a building. The building is 8 times as tall as John. John is 16 years old. John is 5 feet tall. How tall is the building?

Nearly every student with whom we worked used an approach to solving word problems that was mechanical and procedural, rather than based on an attempt to understand the problem. For example, a typical answer for the first word problem noted above was 8 + 9 + 4 = 21. The following explanation about solution strategies was quite typical:

Interviewer: Why did you decide to add the numbers?
Student: Because it said, like, "How far away is the camp?" "How" is to add.

Interviews relevant to the second problem also involved a search for key words in the problems. For example, one student (who was quite typical) stated,

Student: I saw the building is 8 *times* as tall as John so I know to multiply.
Interviewer: What did you multiply?
Student: 16 and 5 and 8.

Our experiments suggested to us that what "mathematical thinking" meant to these students was using procedures to solve numerical problems. The procedures involved a search for key words that specified the operations to perform on the numbers (i.e., add, subtract, multiply, or divide). The numbers to be operated upon were rarely attached to meaningful elements of the problem context. For example, both of the problems noted above include numerical information that was clearly irrelevant (i.e., the fact that 8 other children rode with Tony; that John is 16 years old). Despite this, students consistently attempted to use the irrelevant information in every problem we gave them. Basically, students demonstrated extremely poor comprehension of the problems they were being asked to solve.

Similar findings have been reported by other investigators. For example, several investigators have shown that instead of bringing real-world standards to their work, students seem to treat word problems mechanically and often fail to think about constraints imposed by real-world experiences (Charles & Silver, 1988; Silver, 1986). For example, Silver asked students to determine the number of buses needed to take a specific number of people on a field trip. Many of them divided the total number of students by the number that each bus would hold and came up with answers like $2^{1/3}$. The students failed to consider the fact that one cannot have a functioning third of a bus.

These examples illustrate what has been termed the *inert knowledge* problem (Whitehead, 1929): Knowledge is accessed only in a restricted set of contexts, even though it is applicable to a wide variety of domains. For knowledge to be useful, it must be activated when needed and at the appropriate time (Bransford, Sherwood, Vye, & Rieser, 1986). As discussed, students must understand how procedures can function as tools for solving relevant problems—an understanding that depends on interactive relationships among declarative, procedural, and conceptual knowledge. Environments that promote the development of these relationships should result in knowledge representations that are organized with respect to the *triggering conditions* specifying the applicability of various ideas, facts, procedures, concepts, and so forth. The absence of such conditionalized knowledge (Anderson, 1982, 1987) results in students' not recognizing situations in which their knowledge applies (or, if they do, they may be very slow to solve problems).

Lesh (1981) proposed that the ability to use a mathematical idea depends on the way it is linked to other ideas and processes within an appropriate conceptual model—a cognitive structure that integrates a set of ideas with a system of processes. Difficulties arise because students' mathematical knowledge either is composed of isolated chunks of knowledge or it is linked to conceptual models unconnected with the mathematics in the current problem. Students who have acquired specific knowledge, or even isolated problem-solving strategies, may not have learned to identify (a) situations in which particular strategies might be useful, (b) stages in the problem-solving process when particular strategies might be useful, or (c) relationships among various ideas and strategies. Some cognitive-strategy training attempts to teach children to look for these relationships (see Montague, chapter 9 of this book). Indeed, strategy training is one area in which we are beginning to see the use of hypermedia technology to support children's use of strategic and systematic approaches to simple word problems (Babbitt & Miller, 1996).

In summary, mathematical literacy involves far more than fluent retrieval of basic math facts and the execution of computational procedures. To be sure, learners must possess a sufficient and efficient knowledge base from which to draw information. The retrieval of facts and basic procedural steps necessary to solve a problem depends on recognizing their appropriate use—a skill that requires the coordination of relationships among declarative, procedural, and conceptual knowledge. As we elaborate below, the ability to recognize situations and perform the needed mental elaborations can be strengthened by providing practice in real-life situations involving meaningful and purposeful tasks.

MEANINGFUL LEARNING ENVIRONMENTS FOR MATHEMATICS INSTRUCTION

Our emphasis on meaningful learning environments for mathematics instruction is supported by cognitive research on learning. Cognitive research on transfer-appropriate processing (Bransford, Franks, Morris, & Stein, 1979; Morris, Bransford, & Franks, 1979) indicates that in order to achieve particular outcomes, students must have the opportunity to engage in the kinds of activities that support those outcomes. The implication for mathematics learning is that if we want students to use mathematics to solve complex mathematical problems that arise in day-to-day life, they need opportunities to learn in these contexts. Meaningful problems drawn from everyday contexts are one way to use informal knowledge as a basis for making strong connections between school mathematics and students' informal knowledge (e.g., Ginsburg, chapter 2 of this book; Peterson et al., 1991). Research on motivation also supports the use of meaningful contexts. Several researchers have found that authentic and meaningful tasks are often motivating to students, even if the tasks require a lot of work (Blumenfeld et al., 1991; Bransford et al., 1996; A. Brown & Campione, 1994; Collins, Hawkins, & Carver, 1991; CTGV, 1992c, 1993b, 1994b). (See CTGV, 1996a, 1996b, in press-a, and in press-b, for additional data in support of meaningful learning environments.)

Our approach to the use of meaningful contexts for mathematics instruction has slowly evolved over the past decade. We call the approach *anchored instruction* (CTGV, 1990; 1993a). A complete history of the evolution of anchored instruction is reported in CTGV (in press-a), and a more abbreviated one appears in CTGV (1994a). Briefly, the anchored instruction approach creates meaningful contexts for using content area skills to solve authentic problems. In the case of mathematics, we have created *The Adventures of Jasper Woodbury*, a video-based series that covers four content areas in mathematics (trip planning, statistics and business planning, geometry, and algebra). Each adventure is a narrative that concludes with a challenge in the form of a problem that the student viewers need to solve for the video character(s). The video is recorded on random, accessible storage media, such as laserdiscs or CD ROMs, that allow students to revisit selected parts of the video quickly and easily.

Jasper problems are quite complex. As a result, multiple perspectives are helpful, and often necessary, for solution. Thus, solving anchored instruction problems fosters communication, problem solving, and collaboration. There are numerous reports of the learning that occurs when students work in these environments, and the positive effects occur across the entire range of the achievement spectrum (e.g., CTGV, 1992c, 1994a, in press-a, in press-b; Pellegrino et al., 1991). For the present purposes, what is important is that when we first introduced anchored instruction and the *Jasper* adventures to teachers, many were skeptical and believed that it would be too complex for their "slower" students. This has not turned out to be the case; indeed, a number of these same teachers report that working with *Jasper* has had a powerful, positive impact, especially on those students who traditionally have not done well in mathematics.

A Contrast with Traditional Approaches

The anchored instruction concept frequently, but only initially, meets with resistance among special educators. This is because the idea of learning in complex problem-solving environments runs counter to the recommendations of the behaviorist learning theories that have dominated instructional approaches to students with learning disabilities (for a review, see Rivera, chapter 1 of this book). For example, Gagné's (1968) cumulative learning model was a direct outgrowth of behaviorist assumptions. That model takes a componential and incremental view of learning: Any task is broken down into its components, and each component is taught in isolation from others. As components are mastered, they can be "put together" with other mastered components until all components of the task are learned. At that point, theoretically, the learner should be able to execute the complete task.

However, there are several conceptual flaws with the learning hierarchy model. First, many tasks reflect complex interactions of their components, and, when "put together," the functionalities of the components may fail to resemble those applicable in isolation. In other words, componential analysis and instruction can frequently result in mastery of skills that bear only a slight resemblance to the performance skills needed for the whole task. A second and related problem with the componential approach is that the learner spends lots of time trying to master a skill that may be irrelevant or different in the context of the complete task. Slowness or inability to master the isolated skill may prevent the learner from being presented with the opportunity to engage in the complete task. This has two negative consequences for learners: They never get to see the complete task and understand where the skills fit in, and they are prevented from attempting the complete task because it is assumed that they do not have the incrementally developed skills to perform the task. This very often happens in school tasks. The research literature on writing by students with learning disabilities is a good example. For years, students with learning disabilities were never asked to write essays because they were still mastering sentence-level mechanics. Yet, several research studies indicate that when asked to write about complex ideas, students with learning disabilities often demonstrate conceptual performance that far exceeds what would be predicted based on their performance on lower level skills, such as capitalization, punctuation, and spelling (Graham & Harris, 1989, 1994).

Finally, isolated skill instruction, even when it successfully develops fluent retrieval of declarative knowledge and efficient execution of specific procedures, fails to establish relationships among declarative, procedural, and conceptual knowledge. Hence, it does little to rectify the inert knowledge problem; indeed, it may foster inert knowledge. It is often argued that standard word problems are the vehicle for integrating individual mathematics skills and are the mechanism for including mathematics problem solving in the curriculum (Babbitt & Miller, 1996). However, standard word problems (like those that appear at the end of textbook chapters and are periodically assigned as homework) do not fit the definition of "problem" (Schoenfeld, 1989). It is doubtful that their generic and artificial nature are of much real interest to students because they look nothing like real-world problems. They are often structured in ways that require a simple manipulation of numbers that eventually leads to a correct answer. This is quite different from meaningful and authentic situations encountered in real-world settings—they rarely furnish such hints. In addition, teachers often teach students simply to find "key" words in standard problems, and these key words trigger specific arithmetic operations. These types of shortcuts to solving word problems may actually hinder learners' attempts to more firmly establish the link between arithmetic computation and mathematical understanding. This form of problem-solving exercise does not require a real understanding of the problem situation and relegates the activity to an imitation of drill and practice (Porter, 1989).

Other Examples of Anchors

The *Jasper* episodes are just one example of the kinds of materials that can be used as anchors. An important design feature of anchors is that they provide a motivating and realistic context for problem posing, problem solving, and reasoning. They need to provide students with opportunities to link declarative, procedural, and conceptual knowledge in the context of real-world problems rather than in the unidimensional context of text word problems. As discussed above, the difficulty in teaching students how to solve problems can be attributed, in part, to students' inability to perceive instances in which the knowledge they already possess is useful. The ability to literally "notice" and retrieve useful information appears to be especially problematic for children with learning problems or those who are at risk of school failure and these skills are not developed in traditional word-problem formats (Hasselbring, Goin, Alcantara, & Bransford, 1991).

The deficiencies of word problems are not present in anchored environments. Anchored environments embed problems within a rich video format from which students seek to cull relevant information. The motivating nature of the problems and characters within a story challenge students to work to solve the problems, even though it may take several days to reach a solution. Students identify with the characters in the story and thus are situated in the problem. Finally, the generative nature of the problems requires students to recognize relevant information, formulate strategies, and incorporate mathematical procedures to reach a resolution.

An Anchored Learning Environment for Students with Learning Disabilities

Although the *Jasper* adventures have been used with students across the achievement spectrum, we have also developed a set of prototype anchors specifically for use with students with learning disabilities in mathematics (Bottge & Hasselbring, 1993). We use one of these anchors, called "Ben's Pet Project," to illustrate the design and learning principles we have been describing.* "Ben's Pet Project" opens with Ben sitting on his bed reading a book about reptiles. In the background, a voice from the television can be heard advertising a lumber sale and quoting prices for wood. Ben spreads his money on the bed and counts it. Ben's friend Jimmy comes over to visit and they decide to go to a local pet store to learn whether Ben has enough money to buy a pet like a large rat, a snake, or an iguana. While at the store, they examine a brochure that describes how they can build their own mammal cage. The next scene shows the boys in Ben's garage studying the cage plan and measuring two pieces of wood to be used for constructing the cage shown in the plan. From information given in the video, students are asked how the boys can buy one of the pets in the store and still have enough money left over for a cage.

To solve the overarching problem, students must first gather the relevant facts. All factual information is found in the video. These facts appear in the video in several different ways, all of which parallel natural settings. For example, the cost of lumber is revealed in an advertisement by an announcer on television. The viewer must figure out the amount of money

Ben has to spend from a scene in which he is shown silently counting bills and coins on his bed. Pet prices are shown on the sides of pet cages filmed on location in a pet store, and the pet cage dimensions are detailed in a "do-it-yourself" cage–plan brochure supplied by the pet store.

As in everyday situations, not all of the facts supplied by the video are pertinent to solving the problem. Thus, students must actively engage in a sifting process that separates relevant from irrelevant information. Also, students must define the subproblems themselves in order to reach a resolution to the larger problem that Ben faces. The solution requires that the students be able to use procedural knowledge in several areas, including money, measurement, whole numbers, and fractions. More important, they must understand where and how procedural knowledge in each of these areas is useful and how it is applied.

Bottge and Hasselbring (1993) examined the effectiveness of "Ben's Pet Project" with two groups of adolescents with learning difficulties in mathematics. The two groups were compared on their ability to generate solutions to an anchored transfer problem after being taught problem-solving skills under two conditions. Students in one group were taught problem-solving skills with standard word problems; the second group received instruction using "Ben's Pet Project." The problems presented to both groups focused on procedural knowledge related to money and linear measurement and required students to be able to add and subtract fractions. Students attempted two different transfer tasks following instruction, one being a word-problem analog of the money–linear measurement–fractions problem. The second, an anchored video transfer task, required the use of procedural knowledge in linear measurement, time, money, and decimals for solution.

Results of the study indicated that both groups improved their performance on solving word problems, but that the group instructed using "Ben's Pet Project" did significantly better on the anchored posttest problem. Moreover, unlike students in the word-problem group, students in the anchored environment group were able to transfer skills learned during instruction to textual word-problem and anchored video transfer tasks. We believe that this indicates that much more robust relationships among these students' declarative, procedural, and conceptual knowledge were developed.

Hybrid Model: Skills in the Context of Meaningful and Authentic Tasks

Anchors such as "Ben's Pet Project" and the *Jasper* adventures clearly require that students use "basic" mathematics skills to solve problems.

Authentic problems such as these serve to anchor, tie together, and generate the need-to-know skills, such as multiplication facts, procedures for determining the height of objects, and so forth. Skills taught in the context of meaningful and authentic tasks represent a hybrid approach to mathematics for students with learning disabilities. Technology can assist in the development of hybrid instructional approaches in a number of ways. First, when well-designed computer-based and video-based instruction is provided on a daily basis, children with learning problems can develop declarative and procedural knowledge in basic math skills equivalent to that of their nondisabled peers (Goldman, Mertz, & Pellegrino, 1989; Goldman et al., 1988; Hasselbring et al., 1988; Hasselbring, Sherwood et al., 1991). We envision software environments that help students link skills to particular aspects of authentic tasks and provide support for developing greater fluency with those skills (see CTGV, 1992a).

For example, in one study we conducted, students solved the *Jasper* adventure that involves constructing a business plan to have a dunking booth at the school's upcoming fun fair ("The Big Splash"). Students were quite capable of generating complex plans for solving the problem but were stymied when it came to performing calculations to determine how much it would cost for water if each truckload cost $15 and they needed three truckloads. However, in the context of the business plan, they were highly motivated to figure out how to determine the cost for the water. "Just-in-time" tools (software or paper and pencil) could be linked to this aspect of the problem solving and provide support for students to work on the concepts and procedures surrounding multiplication skills.

Additional work that members of our group at the Learning Technology Center have conducted in the context of the SMART project (Science and Mathematics Arenas for Refining Thinking) has shown the value of solving anchor problems using cycles of feedback and revision (e.g., Barron et al., 1995; CTGV, in press-a). To make this approach manageable for the classroom teacher, we have co-developed with them scoring rubrics that provide students with specific suggestions regarding aspects of their solutions to anchor problems that they need to work on. The rubrics are in Hypercard stacks and are quite easy for the teachers to fill out and print for each student. The rubrics provide individualized feedback about resources students might consult to assist them in the revision process. Some of these resources are additional video-based materials that we have prepared and that develop certain key mathematical concepts anchored to the problem situation they are working on. In the present context, what is important is that these techniques have been effective with students across the achievement spectrum in full-inclusion classrooms.

Our work on formative assessment models in the context of the SMART project highlights the importance of opportunities for ongoing and frequent assessment, which helps students determine how they are doing with respect to weekly, monthly, and yearly academic goals. The curriculum-based measurement approach (Fuchs, Fuchs, & Hamlett, 1992; Fuchs, Fuchs, Hamlett, & Stecker, 1991) would work well in the context of the hybrid model, which we see as the necessary future of mathematics for students with learning disabilities. Keeping track of individual students' progress over the course of repeated assessments and lengthy periods of time is obviously an additional way in which technology can facilitate the development of competence and confidence in mathematics for students with learning disabilities.

We summarize our ideas about hybrid models by proposing a set of design principles for creating learning environments that support achievement of basic mathematical literacy, as defined in new standards and contemporary theories of learning and mathematics. On the basis of data we and others have collected thus far, we have every reason to believe that the following design principles will support mathematics achievement for students with learning disabilities as well as general education students.

1. Situated, meaningful, and authentic problem contexts that motivate the need for fundamental basic skills;
2. Opportunities for development of self-assessment skills, including opportunities for feedback and revision;
3. Support for the acquisition of conceptual understanding of foundational mathematical concepts;
4. Mechanisms for practicing procedural skills, such as addition, subtraction, formulaic manipulation, and so forth; and
5. Support for developing multiple ways to represent and communicate information.

SUMMARY AND CONCLUSIONS

In this chapter we have argued that it is no longer sufficient for any student to simply master the skills of mathematical computation. National commissions and business and mathematics organizations recommend placing much more emphasis on mathematical thinking and problem solving and spending less time on rote-memory and computation activities. However, missing from much of the discussion about what should be done are the practical methods that general and special education teachers can use to teach problem solving to students with learning disabilities.

We reviewed research suggesting that even when declarative and procedural knowledge are in place, students with learning disabilities often fail to apply that knowledge in meaningful ways when confronted with problem situations. When confronted with a problem, students with learning disabilities typically do not use the mathematical knowledge they have acquired to solve problems unless they are explicitly informed about the relationship between the knowledge and the problem. For knowledge to be useful, students must understand how procedures can function as tools for solving relevant problems.

We have argued that current modes of mathematics instruction rarely address the issue of inert knowledge. Standard word problems that appear at the end of textbook chapters and are periodically assigned as homework do not provide students with learning disabilities with an opportunity to understand how mathematical information can be used to solve real-world problems. This form of problem-solving exercise does not require a real understanding of the problem situation and fails to help students develop the linkages between procedural and conceptual knowledge that they need.

Thus, as educators, we are faced with a problem: How can we supply problem-solving situations in contexts that students with learning disabilities view as important? Is it possible to arrange the environment so that patterns of imported knowledge structures match their anticipated uses? NAEP assessments make it apparent that answers to these questions have been, and continue to be, elusive.

We discussed as a possible solution to this problem the merger of technology and cognitive learning theory into an instructional method called anchored instruction. In our approach, cognitive theory forms the conceptual base for what is delivered, and technology provides the tool for delivering it. Anchored instruction attempts to overcome inert knowledge and make it useful for solving problems across unique situations. These problem-solving strategies are gradually linked into webs of knowledge that are more quickly retrieved as situations for their use are recognized. The authenticity of the problems makes them real, not tricky, as is the case with standard word problems. Students are motivated to solve new problems when the problems feel real to them. Other chapters in this book suggest additional solutions that would be consistent with the approach we advocate (e.g., Carnine; Ginsburg; Thornton, Langrall, & Jones).

Technology serves as the delivery medium. One approach to delivering anchored instruction is through the use of short video vignettes in the form of contextualized problems. Interactive video technology allows students and the teacher to instantly access sections of the video for review. A carefully constructed contextualized problem requires students to notice what is important for solving the problem and gives them practice in organizing individual bits of information into strategies for solving problems.

As discussed, the successful use of video anchors with students who have experienced difficulty in mathematics is encouraging. These anchored environments provide students with experience in using math skills to solve real-world problems. Educators have been challenged to find ways of improving the problem-solving skills of all students. Anchored learning environments may provide an effective, practical way of meeting that challenge. As discussed in this chapter, even the performance of students experiencing serious problems in mathematics can be improved by instruction focusing on video problems mediated by the teacher. Cognitive learning theory and technology finally may have merged to provide a powerful teaching/learning tool for educators looking for a practical way of fostering students' problem-solving skills.

Authors' Note

Members of the Cognition and Technology Group at Vanderbilt contributing to the Center's work in mathematics are (in alphabetical order) Brigid Barron, John Bransford, Trefor Davies, Laura Goin, Allison Moore, Priscilla Moore, Tom Noser, James Pellegrino, Daniel Schwartz, Nancy Vye, and Linda Zech.

References

Foreword

Jitendra, A. K., & Xin, Y. P. (1997). *Mathematics word-problem solving instruction for students with mild disabilities and students at risk for math failure: A research synthesis. Journal of Special Education, 30, 412–438.*

Lessen, E., Dudzinski, M., Karsh, K., & Van Acker, R. (1989). *A survey of 10 years of academic intervention research with learning disabled students: Implications for research and practice. Learning Disabilities Focus, 4(2), 106–122.*

Mastropieri, M. A., Scruggs, T. E., & Shiah, S. (1991). *Mathematics instruction for learning disabled students: A review of research. Learning Disabilities Research & Practice, 6, 89–98.*

McLesky, J., Waldron, N. L. (1990). *The identification and characteristics of students with learning disabilities in Indiana. Learning Disabilities Research 5(2), 72–78.*

National Council for Teachers of Mathematics. (1989). *Curriculum and evaluation standards for school mathematics. Reston, VA: Author. (ERIC Document Reproduction Service No. ED 304 336)*

National Research Council. (1989). *Everybody counts: A report to the nation on the future of mathematics education. Washington, DC: National Academy Press.*

Chapter 1

Adelman, P. B., & Vogel, S. A. (1991). *The learning-disabled adult. In B. Y. L. Wong (Ed.), Learning about learning disabilities (pp. 564–594). San Diego: Academic Press.*

Beirne-Smith, M. (1991). *Peer tutoring in arithmetic for children with learning disabilities. Exceptional Children, 57, 330–337.*

Blankenship, C. S. (1978). *Remediating systematic inversion errors in subtraction through the use of demonstration and feedback. Learning Disability Quarterly, 1, 12–22.*

Blankenship, C. S., & Baumgartner, M. D. (1982). *Programming generalization of computational skills. Learning Disability Quarterly, 5, 152–162.*

Bley, N. S., & Thornton, C. A. (1995). Teaching mathematics to students with learning disabilities. Austin, TX: PRO-ED.

Bloom, B. S. (Ed.). (1956). Taxonomy of educational objectives. New York: McKay.

Bottge, B. A., & Hasselbring, T. S. (1993). A comparison of two approaches for teaching complex, authentic mathematics problems to adolescents in remedial math classes. Exceptional Children, 59, 556–566.

Carnine, D., Engelmann, S., Hofmeister, A., & Kelly, B. (1987). Videodisc instruction in fractions. Focus on Learning Problems in Mathematics, 9(1), 31–52.

Carpenter, T. P., Coburn, T. G., Reyes, R. E., & Wilson, J. W. (1976). Notes from the national assessment: Problem solving. Mathematics Teacher, 32, 389–393.

Carpenter, T. P., Corbitt, M. K., Kepner, H. S., Jr., Lindquist, M. M., & Reyes, R. E. (1981). Results from the second mathematics assessment of the national assessment of educational progress. Reston, VA: National Council of Teachers of Mathematics.

Case, L. P., & Harris, K. R. (1988, April). Self-instructional strategy training: Improving mathematical problem solving skills of learning disabled students. Paper presented at the annual meeting of the American Educational Research Association, New Orleans.

Case, L. P., Harris, K. R., & Graham, S. (1992). Improving the mathematical problem-solving skills of students with learning disabilities: Self-regulated strategy development. The Journal of Special Education, 26, 1–19.

Cawley, J. F., Baker-Kroczynski, S., & Urban, A. (1992). Seeking excellence in mathematics education for students with mild disabilities. Teaching Exceptional Children, 24, 40–43.

Cawley, J. F., & Miller, J. H. (1989). Cross-sectional comparisons of the mathematical performance of children with learning disabilities: Are we on the right track toward comprehensive programming? Journal of Learning Disabilities, 23, 250–254, 259.

Cawley, J. F., & Parmar, R. S. (1992). Arithmetic programming for students with disabilities: An alternative. Remedial and Special Education, 13(3), 6–18.

Chiang, B. (1986). Initial learning and transfer effects of microcomputer drills on LD students' multiplication skills. Learning Disability Quarterly, 9, 118–123.

Cobb, P., Yackel, E., & Wood, T. (1992). A constructivist alternative to the representational view of mind in mathematics education. Journal for Research in Mathematics Education, 23, 2–33.

Collins, M., & Carnine, D. (1988). Evaluating the field test revision process by comparing two versions of a reasoning skills CAI program. Journal of Learning Disabilities, 21, 375–379.

Conference Board of the Mathematical Sciences. (1995, May). Adopted statement at CBMS Council meeting. Washington, DC: CBMS.

Cooney, T. J. (1994). Research and teacher education: In search of common ground. Journal for Research in Mathematics Education, 25, 608–636.

Cybriwsky, C. A., & Schuster, J. W. (1990). Using constant delay procedures to teach multiplication facts. Remedial and Special Education, 11(1), 54–59.

Dangel, H. L., & Ensminger, E. E. (1988). The use of a discrepancy formula with LD students. Learning Disabilities Focus, 4(1), 24–31.

Dossey, J., Mullis, I., Lindquist, M., & Chambers, D. (1988). *The mathematics report card: Are we measuring up?* Princeton, NJ: Educational Testing Service.

Driscoll, M. (1983). *Research within reach: Secondary school mathematics.* Reston, VA: National Council of Teachers of Mathematics.

Englert, C. S., Culatta, B. E., & Horn, D. G. (1987). Influence of irrelevant information in addition word problems on problem solving. *Learning Disability Quarterly, 10,* 29–36.

Epps, S., Ysseldyke, J. E., & McGue, M. (1984). "I know one when I see one"—Differentiating LD from non-LD students. *Learning Disability Quarterly, 7,* 89—101.

Fennema, E., Carpenter, T. P., & Lamon, S. J. (Eds.). (1988). *Integrating research on teaching and learning mathematics.* Albany: State University of New York Press.

Feuer, M. J., & Fulton, K. (1993). The many faces of performance assessment. *Phi Delta Kappan, 74(6),* 478.

Frame, R. E., Clarizio, H. F., & Porter, A. (1984). Diagnostic and prescriptive bias in school psychologists' reports of a learning disabled child. *Journal of Learning Disabilities, 17,* 12–15.

Friedman, S. G., & Hofmeister, A. M. (1984). Matching technology to content and learners: A case study. *Exceptional Children, 51,* 130–134.

Fuchs, L. S., Bahr, C. M., & Rieth, H. J. (1989). Effects of goal structures and performance contingencies on math performance of adolescents with learning disabilities. *Journal of Learning Disabilities, 22,* 554–560.

Fuchs, L. S., Fuchs, D., & Hamlett, C. L. (1989). Computers and curriculum-based measurement: Effects of teacher feedback systems. *School Psychology Review, 18,* 112–125.

Fuchs, L. S., Fuchs, D., Hamlett, C. L., & Stecker, P. M. (1990). The contribution of skills analysis to curriculum-based measurement in math. *School Psychology Review, 19,* 6–22.

Fuchs, L. S., Fuchs, D., Hamlett, C. L., & Stecker, P. M. (1991). Effects of curriculum-based measurement and consultation on teacher planning and student achievement in mathematics operations. *American Educational Research Journal, 28,* 617–641.

Gagné, R. M. (Ed.). (1962). *Psychological principles in system development.* New York: Holt, Rinehart & Winston.

Garnett, K. (1987). Math learning disabilities: Teaching and learners. *Reading, Writing, and Learning Disabilities, 3,* 1–8.

Garnett, K., & Fleischner, J. E. (1983). Automatization and basic fact performance of normal and learning disabled children. *Learning Disability Quarterly, 6,* 223–231.

Geary, D. C. (1990). A componential analysis of an early learning deficit in mathematics. *Journal of Experimental Child Psychology, 49,* 363–383.

Geary, D. C. (1993). Mathematical disabilities: Cognitive, neuropsychological, and genetic components. *Psychological Bulletin, 114,* 345–362.

Gersten, R., Carnine, D., & Woodward, J. (1987). Direct instruction research: The third decade. *Remedial and Special Education, 8(6),* 48–56.

Ginsburg, H. P. (1989). *Children's arithmetic: How they learn it and how you teach it* (2nd ed.). Austin, TX: PRO-ED.

Gleason, M., Carnine, D., & Boriero, D. (1990). Improving CAI effectiveness with attention to instructional design in teaching story problems to mildly handicapped students. *Journal of Special Education Technology, 10(3),* 129–136.

Goldin, G. A., & McClintock, C. E. (Eds.). (1984). *Task variables in mathematical problem solving.* Hillsdale, NJ: Erlbaum.

Goldman, S. R., Pellegrino, J. W., & Mertz, D. L. (1988). Extended practice of basic addition facts: Strategy changes in learning disabled students. *Cognition & Instruction, 5,* 223–265.

Graves, A., Landers, M. F., Lokerson, J., Luchow, J., Horvath, M., & Garnett, K. (1992). *The DLD competencies for teachers of students with learning disabilities.* Reston, VA: Division for Learning Disabilities, Council for Exceptional Children.

Hammill, D. D. (Ed.). (1987). *Assessing the abilities and instructional needs of students.* Austin, TX: PRO-ED.

Harding, D., Gust, A. M., Goldhawk, S. L., & Bierman, M. M. (1993). The effects of the Interactive Unit on the computation skills of students with learning disabilities and students with mild cognitive impairments. *Learning Disabilities: A Multidisciplinary Journal, 4(2),* 53–65.

Hécaen, H., Angelergues, R., & Houillier, S. (1961). Les variétiés clinique des acalculies au cours des lésions rétrorolandiques: Approche statistique du problème [The clinical varieties of the acalculies in retrorolandic lesions: A statistical approach to the problem]. *Revue Neurologique, 105,* 85–103.

Hofmeister, A. M. (1993). Elitism and reform in school mathematics. *Remedial and Special Education, 14(6),* 8–13.

Howell, R., Sidorenko, E., & Jurica, J. (1987). The effects of computer use on the acquisition of multiplication facts by a student with learning disabilities. *Journal of Learning Disabilities, 20,* 336–341.

Hutchinson, N. L. (1993a). Students with disabilities and mathematics education reform—Let the dialogue begin. *Remedial and Special Education, 14(6),* 20–23.

Hutchinson, N. L. (1993b). The effect of cognitive instruction on algebra problem solving of adolescents with learning disabilities. *Learning Disability Quarterly, 16,* 34–63.

Johnson, D. C., Romberg, T. A., & Scandura, J. M. (1994). The origins of the JRME: A retrospective account. *Journal for Research in Mathematics Education, 25(6),* 561–582.

Johnston, W. B., & Packers, A. E. (1987). *Workforce 2000: Work and workers for the twenty-first century.* Indianapolis: Hudson Insitute.

Jones, G. A., Thornton, C. A., & Toohey, M. A. (1985). A multiplication program for learning basic addition facts: Case studies and an experimental report. *Journal of Learning Disabilities, 18,* 319–325.

Kane, B. J., & Alley, G. R. (1980). A peer tutored instructional management program in computational mathematics for incarcerated learning disabled juvenile delinquents. *Journal of Learning Disabilities, 13,* 39–42.

Kelly, B., Carnine, D., Gersten, R., & Grossen, B. (1986). The effectiveness of videodisc instruction in teaching fractions to learning-disabled and remedial high school students. Journal of Special Education Technology, 8(2), 5–17.

Kelly, B., Gersten, R., & Carnine, D. (1990). Student error patterns as a function of curricular design: Teaching fractions to remedial high school students and high school students with learning disabilities. Journal of Learning Disabilities, 23, 23–29.

Kieran, C. (1994). Doing and seeing things differently: A 25-year retrospective of mathematics education research on learning. Journal for Research in Mathematics Education, 25, 583–607.

Kilpatrick, J. (1992). A history of research in mathematics education. In D. Grouws (Ed.), Handbook for research on mathematics teaching and learning (pp. 3–38). New York: Macmillan.

Kosc, L. (1974). Developmental dyscalculia. Journal of Learning Disabilities, 7, 165–177.

Koscinski, S. T., & Gast, D. L. (1993). Computer-assisted instruction with constant time delay to teach multiplication facts to students with learning disabilities. Learning Disabilities Research & Practice, 8(3), 157–168.

Leon, J. A., & Pepe, H. J. (1983). Self-instructional training: Cognitive behavior modification for remediating arithmetic deficits. Exceptional Children, 50, 54–60.

Lessen, E., Dudzinski, M., Karsh, K., & Van Acker, R. (1989). A survey of ten years of academic intervention research with learning disabled students: Implications for research and practice. Learning Disabilities Focus, 4(2), 106–122.

Lester, F. K., Jr. (1994). Musings about mathematical problem-solving research: 1970–1994. Journal for Research in Mathematics Education, 25, 660–675.

Lester, F. K., Jr., & Garofalo, J. (Eds.). (1982). Mathematical problem solving: Issues in research. Philadelphia: Franklin Institute.

Lindquist, M., Carpenter, T. P., Silver, E., & Mathews, W. (1983). The third national mathematics assessment: Results and implications for elementary and middle schools. The Arithmetic Teacher, 31(4), 14–19.

Lloyd, J., Saltzman, N. J., & Kauffman, J. M. (1981). Predictable generalization in academic learning as a result of preskills and strategy training. Learning Disability Quarterly, 4, 203–216.

Luiselli, J. K., & Downing, J. N. (1980). Improving a student's arithmetic performance using feedback and reinforcement procedures. Education and Treatment of Children, 3, 45–49.

Maheady, L., Sacca, M. K., & Harper, G. F. (1987). Classwide student tutoring teams: The effects of peer-mediated instruction on the academic performance of secondary mainstreamed students. The Journal of Special Education, 21, 107–121.

Marzola, E. (1985). An arithmetic problem solving model based on a plan for steps to solution, mastery learning, and calculator use in a resource room setting for learning disabled students. New York: Columbia University Teachers College.

Mastropieri, M. A., Scruggs, T. E., & Shiah, S. (1991). Mathematics instruction for learning disabled students: A review of research. Learning Disabilities Research & Practice, 6, 89–98.

McIntyre, S. B., Test, D. W., Cooke, N. L., & Beattie, J. (1991). Using count-bys to increase multiplication facts fluency. Learning Disability Quarterly, 14, 82–88.

McKnight, C., Crosswhite, F., Dossey, J., Kifer, E., Swafford, J., Travers, K., & Cooney, T. (1987). *The underachieving curriculum: Assessing U.S. school mathematics from an international perspective.* Champaign, IL: Stipes.

McLeod, T., & Armstrong, S. (1982). Learning disabilities in mathematics—Skill deficits and remedial approaches. *Learning Disability Quarterly, 5,* 305–311.

McLoughlin, J. A., & Lewis, R. B. (1990). *Assessing special students.* New York: Macmillan.

Mercer, C. D., Harris, C. A., & Miller, S. P. (1993). Reforming reforms in mathematics. *Remedial and Special Education, 14(6),* 14–19.

Mercer, C. D., & Miller, S. P. (1992). Teaching students with learning problems in math to acquire, understand, and apply basic math facts. *Remedial and Special Education, 13(3),* 19–35, 61.

Miller, S. P., & Mercer, C. D. (1993). Mnemonics: Enhancing the math performance of students with learning difficulties. *Intervention in School and Clinic, 29,* 78-82.

Montague, M. (1992). The effects of cognitive and metacognitive strategy instruction on the mathematical problem solving of middle school students with learning disabilities. *Journal of Learning Disabilities, 25,* 230–248.

Montague, M., & Applegate, B. (1993). Middle school students' mathematical problem solving: An analysis of think-aloud protocols. *Learning Disability Quarterly, 16,* 19–30.

Montague, M., & Bos, C. S. (1986). The effect of cognitive strategy training on verbal math problem solving performance of learning disabled adolescents. *Journal of Learning Disabilities, 19,* 26–33.

National Commission on Excellence in Education. *A nation at risk: The imperative for education reform.* Washington, DC: U.S. Government Printing Office.

National Council of Supervisors of Mathematics. (1988). *Twelve components of essential mathematics.* Minneapolis: Author.

National Council of Teachers of Mathematics. (1980). *An agenda for action.* Reston, VA: Author.

National Council of Teachers of Mathematics. (1989). *Curriculum and evaluation standards for school mathematics.* Reston, VA: Author.

National Council of Teachers of Mathematics. (1991). *Professional standards for teaching mathematics.* Reston, VA: Author.

National Research Council. (1989). *Everybody counts—A report to the nation on the future of mathematics education.* Washington, DC: National Academy.

National Science Board Commission. (1983). *Educating Americans for the 21st century.* Washington, DC: National Science Foundation.

Office of Technology Assessment. (1988). *Technology and the American transition.* Washington, DC: U.S. Government Printing Office.

Parmar, R. S. (1992). Protocol analysis of strategies used by students with mild disabilities when solving arithmetic word problems. *Diagnostique, 17(4),* 227–243.

Parmar, R. S., Cawley, J. F., & Frazita, R. R. (1996). Word problem-solving by students with and without mild disabilities. *Exceptional Children, 62,* 415–429.

Peterson, S. K., Mercer, C. D., & O'Shea, L. (1988). Teaching learning disabled students place value using the concrete to abstract sequence. *Learning Disabilities Research, 4(1),* 52–56.

Phillips, N. B., Hamlett, C. L., Fuchs, L. S., & Fuchs, D. (1993). Combining classwide curriculum-based measurement and peer tutoring to help general educators provide adaptive education. Learning Disabilities Research & Practice, 8(3), 148–156.

Piaget, J. (1954). The construction of reality in the child (M. Cook, Trans.). New York: Basic Books.

Piaget, J. (1963). Origins of intelligence in children. New York: Norton.

Piaget, J. (1970). The science of education and the psychology of the child. New York: Orion.

Poplin, M., Wiest, D., & Thorson, S. (1995). Alternative instructional strategies to reductionism: Constructive, multicultural, feminine and critical pedagogies. In W. Stainback & S. Stainback (Eds.), Controversial issues confronting special education: Divergent perspectives (2nd ed.). Boston: Allyn & Bacon.

Reynolds, C. (1985). Measuring the aptitude–achievement discrepancy in learning disability diagnosis. Remedial and Special Education, 6, 37–55.

Rivera, D. (1993). Examining mathematics reform and the implications for students with mathematical disabilities. Remedial and Special Education, 14(6), 24–27.

Rivera, D. P., Carter, A., & Smith, R. (1996). The effects of computer-based instruction and fluency building techniques on the proficiency and accuracy of answering math facts by students with learning disabilities. Unpublished manuscript.

Rivera, D., & Smith, D. D. (1987). Influence of modeling on acquisition and generalization of computational skills: A summary of research findings from three sites. Learning Disability Quarterly, 10, 69–80.

Rivera, D., & Smith, D. D. (1988). Using a demonstration strategy to teach learning disabled midschool students how to compute long division. Journal of Learning Disabilities, 21, 77–81.

Rivera, D. P., Taylor, R. L., & Bryant, B. R. (1994–95). A review of current trends in mathematics assessment for students with mild disabilities. Diagnostique, 20, 143–174.

Scheid, K. (1990). Cognitive-based methods for teaching mathematics to students with learning problems. Columbus, OH: Information Center for Special Education Media and Materials.

Schoenfeld, A. H. (1982). Some thoughts on problem-solving research and mathematics education. In F. K. Lester & J. Garofalo (Eds.), Mathematical problem solving: Issues in research (pp. 27–37). Philadelphia: Franklin Institute.

Schoenfeld, A. H. (1985). Mathematical problem solving. Orlando, FL: Academic Press.

Schoenfeld, A. H. (1994). A discourse on methods. Journal for Research in Mathematics Education, 25, 697–710.

Schunk, D. H. (1985). Participation in goal setting: Effects on self-efficacy and skills of learning disabled children. The Journal of Special Education, 19, 307–317.

Schunk, D. H., & Cox, P. D. (1986). Strategy training and attribution feedback with learning disabled students. Journal of Educational Psychology, 78, 201–209.

Skemp, R. R. (1976). Relational understanding and instrumental understanding. Mathematics Teaching, 77, 20–26.

Skinner, B. F. (1968). The technology of teaching. New York: Appleton-Century-Crofts.

Slavin, R. E., Madden, N. A., & Leavey, M. (1984). *Effects of team assisted individu-alization on the mathematics achievement of academically handicapped and non-handicapped students. Journal of Educational Psychology, 76, 813–819.*

Smith, D. D., & Lovitt, T. C. (1975). *The use of modeling techniques to influence the acquisition of computational arithmetic skills in learning-disabled children. In E. Ramp & G. Semb (Eds.), Behavior analysis: Areas of research and application (pp. 283–308). Englewood Cliffs, NJ: Prentice-Hall.*

Smith, D. D., & Lovitt, T. C. (1976). *The differential effects of reinforcement contin-gencies on arithmetic performance. Journal of Learning Disabilities, 9, 32–40.*

Suydam, M. N. (1980). *Untangling clues from research on problem solving. In S. Krulik (Ed.), Problem solving in school mathematics (pp. 34–50). Reston, VA: National Council of Teachers of Mathematics.*

Trifiletti, J. J., Frith, G. H., & Armstrong, S. (1984). *Microcomputers versus resource rooms for LD students: A preliminary investigation of the effects on math skills. Learning Disability Quarterly, 7, 69–76.*

Twentieth Century Fund. (1983). *Making the grade. New York: Author.*

Vygotsky, L. (1978). *Mind in society: The development of higher psychological processes. Cambridge, MA: Harvard University.*

Wagner, M. (1990). *The school programs and school performance of secondary students classified as learning disabled: Findings from the National Longitudinal Transition Study of Special Education Students. Menlo Park, CA: SRI International.*

Wesson, C. L., & King. R. P. (1992). *The role of curriculum-based measurement in port-folio assessment. Diagnostique, 18(1), 27–37.*

Wittrock, M. C. (1974). *A generative model of mathematics learning. Journal for Research in Mathematics Education, 5, 181–196.*

Woodward, J. (1991). *Procedural knowledge in mathematics: The role of the curriculum. Journal of Learning Disabilities, 24, 242–251.*

Worthern, B. R. (1993). *Critical issues that will determine the future of alternative assessment. Phi Delta Kappan, 74, 444–454.*

Zawaiza, T., & Gerber, M. (1993). *Effects of explicit instruction on math word-problem solving by community college students with learning disabilities. Learning Disability Quarterly, 16, 64–79*

CHAPTER 2

Antell, S., & Keating, D. (1983). *Perception of numerical invariance in neonates. Child Development, 54, 695–701.*

Arbeiter, S. (1984). *Profiles, college-bound seniors, 1984. New York: College Entrance Examination Board.*

Badian, N. A. (1983). *Dyscalculia and nonverbal disorders of learning. In H. R. Mykelbust (Ed.), Progress in learning disabilities (pp. 235–264). New York: Stratton.*

Baroody, A. J. (1987). *Children's mathematical thinking. New York: Teachers College Press.*

Binet, A., & Simon, T. (1916). *The development of intelligence in children*. Baltimore: Williams & Wilkins.

Bloom, L. (1970). *Language development: Form and function in emerging grammars*. Cambridge, MA: MIT.

Campione, J. C., Brown, A. L., Ferrara, R. A., & Bryant, N. R. (1984). The zone of proximal development: Implications for individual differences and learning. In B. Rogoff & J. V. Wertsch (Eds.), *New directions for child development: Children's learning in the zone of proximal development* (pp. 77–91). San Francisco: Jossey-Bass.

Carraher, T. N., Carraher, D. W., & Schliemann, A. S. (1985). Mathematics in streets and schools. *British Journal of Developmental Psychology, 3*, 21–29.

Dantzig, T. (1954). *Number: The language of science* (4th ed.). New York: Macmillan.

Eisenhart, M., Borko, H., Underhill, R., Brown, C., Jones, D., & Agard, P. (1993). Conceptual knowledge falls through the cracks: Complexities of learning to teach mathematics for understanding. *Journal for Research in Mathematics Education, 24(1)*, 8–40.

Ericsson, K. A., & Simon, H. A. (1993). *Protocol analysis: Verbal reports as data*. Cambridge, MA: MIT Press.

Farnham-Diggory, S. (1992). *The learning-disabled child*. Cambridge, MA: Harvard University Press.

Ferrini-Mundy, J., & Lauten, D. (1993). Teaching and learning calculus. In P. S. Wilson (Ed.), *Research ideas for the classroom: High school mathematics* (pp. 155–176). New York: Macmillan.

Fetterman, D. (1989). *Ethnography: Step by step*. Newbury Park, CA: Sage.

Geary, D. C. (1990). A componential analysis of an early learning deficit in mathematics. *Journal of Experimental Child Psychology, 49*, 363–383.

Geary, D. C. (1993). Mathematical disabilities: Cognitive, neuropsychological, and genetic components. *Psychological Bulletin, 114*, 345–362.

Gelman, R. (1980). What young children know about numbers. *Educational Psychologist, 15*, 54–68.

Ginsburg, H. P. (1986). The myth of the deprived child: New thoughts on poor children. In U. Neisser (Ed.), *The school achievement of minority children: New perspectives* (pp. 169–189). Hillsdale, NJ: Erlbaum.

Ginsburg, H. P. (1989). *Children's arithmetic: How they learn it and how you teach it* (2nd ed.). Austin, TX: PRO-ED.

Ginsburg, H. P. (1990). *Assessment probes and instructional activities: The test of early mathematics ability* (2nd ed.). Austin, TX: PRO-ED.

Ginsburg, H. P. (in press). Toby's math. In R. J. Sternberg & T. Ben-Zeev (Eds.), *The nature of mathematical thinking*. Hillsdale, NJ: Erlbaum.

Ginsburg, H. P., Jacobs, S. F., & Lopez, L. S. (1993). Assessing mathematical thinking and learning potential. In R. B. Davis & C. S. Maher (Eds.), *Schools, mathematics, and the world of reality* (pp. 237–262). Boston: Allyn & Bacon.

Ginsburg, H. P., Posner, J. K., & Russell, R. L. (1981). The development of mental addition as a function of schooling and culture. *Journal of Cross-Cultural Psychology, 12*, 163–178.

Ginsburg, H. P., & Russell, R. L. (1981). Social class and racial influences on early mathematical thinking. Monographs of the Society for Research in Child Development, 46 (No. 6, Serial No. 193).

Glaser, R. (1981). The future of testing: A research agenda for cognitive psychology and psychometrics. American Psychologist, 36, 923–936.

Greenfield, P. M. (1984). A theory of the teacher in the learning activities of everyday life. In B. Rogoff & J. Lave (Eds.), Everyday cognition: Its development in social context (pp. 95–116). Cambridge, MA: Harvard University Press.

Groen, G., & Resnick, L. B. (1977). Can preschool children invent addition algorithms? Journal of Educational Psychology, 69, 645–652.

Hughes, M. (1986). Children and number: Difficulties in learning mathematics. New York: Basil Blackwell.

Klein, A., & Starkey, P. (1988). Universals in the development of early arithmetic cognition. In G. Saxe & M. Gearhart (Eds.), Children's mathematics (pp. 5–26). San Francisco: Jossey-Bass.

Kozol, J. (1991). Savage inequalities: Children in America's schools. New York: Crown.

Kuhn, D. (1992). Cognitive development. In M. H. Bornstein & M. E. Lamb (Eds.), Developmental psychology: An advanced textbook (pp. 211–272). Hillsdale, NJ: Erlbaum.

Kuhn, D., & Phelps, E. (1982). The development of problem-solving strategies. In H. Reese & L. Lipsitt (Eds.), Advances in child development and behavior (pp. 1–44). San Diego: Academic Press.

McDermott, R. P. (1993). The acquisition of a child by a learning disability. In S. Chaiklin & J. Lave (Eds.), Understanding practice: Perspectives on activity and context (pp. 269–305). Cambridge, England: Cambridge University Press.

National Center for Education Statistics. (1990). Digest of education statistics. Washington, DC: U.S. Department of Education, Office of Educational Research and Improvement.

National Council of Teachers of Mathematics. (1989). Curriculum and evaluation standards for school mathematics. Reston, VA: Author.

Natriello, G., McDill, E. L., & Pallas, A. M. (1990). Schooling disadvantaged children: Racing against catastrophe. New York: Teachers College Press.

Noddings, N. (1990). Constructivism in mathematics education. In R. B. Davis, C. A. Maher, & N. Noddings (Eds.), Constructivist views on the teaching and learning of mathematics (pp. 7–18). Reston, VA: National Council of Teachers of Mathematics.

Nunes, T., Schliemann, A. D., & Carraher, D. W. (1993). Street mathematics and school mathematics. Cambridge, England: Cambridge University.

Piaget, J. (1952a). The child's conception of number (C. Gattegno & F.M. Hodgson, Trans.). London: Routledge & Kegan Paul Ltd. (Original work published 1941)

Piaget, J. (1952b). The origins of intelligence in children (M. Cook, Trans.). New York: International Universities. (Original work published 1936)

Pillemer, D. B., & White, S. H. (1989). Childhood events recalled by children and adults. In H. W. Reese (Ed.), Advances in child development and behavior (Vol. 21, pp. 83–98). New York: Academic Press.

Poplin, M. S. (1988). The reductionistic fallacy in learning disabilities: Replicating the past by reducing the present. Journal of Learning Disabilities, 21, 389–400.

Post, T. R., Harel, G., Behr, M. J., & Lesh, R. (1991). Intermediate teachers' knowledge of rational number concepts. In E. Fennema, T. P. Carpenter, & S. J. Lamon (Eds.), Integrating research on teaching and learning mathematics (pp. 177–198). Albany, NY: SUNY.

Resnick, L. B., Levine, J. M., & Teasley, S. D. (Eds.). (1991). Perspectives on socially shared cognition. Washington, DC: American Psychological Association.

Royer, J. M., Cisero, C. A., & Carlo, M. S. (1993). Techniques and procedures for assessing cognitive skills. Review of Educational Research, 63, 201–243.

Russell, R. L., & Ginsburg, H. P. (1984). Cognitive analysis of children's mathematics difficulties. Cognition and Instruction, 1, 217–244.

Saxe, G., Guberman, S. R., & Gearhart, M. (1987). Social processes in early number development. Monographs of the Society for Research in Child Development, 52 (2, Serial No. 216).

Schoenfeld, A. H. (1985). Making sense of "Out Loud" problem-solving protocols. Journal of Mathematical Behavior, 4, 171–191.

Schoenfeld, A. H. (1989). Exploration of students' mathematical beliefs and behavior. Journal for Research in Mathematics Education, 20, 338–355.

Shaywitz, S. E., Escobar, M. D., Shaywitz, B. A., Fletcher, J. M., & Makuch, R. (1992). Evidence that dyslexia may represent the lower tail of a normal distribution of reading ability. The New England Journal of Medicine, 236, 145–150.

Siegler, R. S. (1988). Strategy choice procedures and the development of multiplication skill. Journal of Experimental Psychology: General, 117, 258–275.

Siegler, R. S., & Crowley, K. (1991). The microgenetic method: A direct means for studying cognitive development. American Psychologist, 46, 606–620.

Smith, N. J., & Wendelin, K. H. (1981). Using children's books to teach mathematical concepts. Arithmetic Teacher, 29, 10–15.

Stevenson, H., Lee, S. S., & Stigler, J. (1986). The mathematics achievement of Chinese, Japanese, and American children. Science, 56, 693–699.

Stevenson, H. W., & Stigler, J. W. (1992). The learning gap: Why our schools are failing and what we can learn from Japanese and Chinese education. New York: Summit.

Stigler, J. W., & Perry, M. (1988). Mathematics learning in Japanese, Chinese, and American classrooms. In G. B. Saxe & M. Gearhart (Eds.), Children's mathematics (pp. 27–54). San Francisco: Jossey-Bass.

Thompson, A. G. (1992). Teachers' beliefs and conceptions: A synthesis of the research. In D. A. Grouws (Ed.), Handbook of research in mathematics teaching and learning (pp. 127–146). New York: Macmillan.

Vygotsky, L. S. (1978). Mind in society: The development of higher psychological processes (M. Cole, Trans.). Cambridge, MA: Harvard University Press. (Original work published 1930)

Vygotsky, L. S. (1986). Thought and language (A. Kozulin, Trans.). Cambridge, MA: MIT Press. (Original work published 1934)

Whitehead, A. N. (1929). The aims of education. New York: Macmillan.

CHAPTER 3

Badian, N. A. (1983). Dyscalculia and nonverbal disorders of learning. In H. R. Myklebust (Ed.), Progress in learning disabilities (Vol. 5, pp. 235–264). New York: Grune & Stratton.

Benson, D. F., & Weir, W. F. (1972). Acalculia: Acquired anarithmetria. Cortex, 8, 465–472.

Benton, A. L. (1961). The fiction of the Gerstmann syndrome. Journal of Neurology, Neurosurgery, and Psychiatry, 24, 176–181.

Benton, A. L. (1987). Mathematical disability and the Gerstmann syndrome. In G. Deloche & X. Seron (Eds.), Mathematical disabilities: A cognitive neuropsychological perspective (pp. 111–120). Hillsdale, NJ: Erlbaum.

Benton, A. L. (1992). Gerstmann's syndrome. Archives of Neurology, 49, 445–447.

Berger, H. (1926). Ueber Rechenstoerungen bei Herderkrankungen des Grosshirns [On arithmetic problems of focal diseases of the cerebrum]. Archiv Fuer Psychiatrie, 78, 238–263.

Boll, T. J., & Barth, J. T. (1981). Neuropsychology of brain damage in children. In S. B. Filskov & T. J. Boll (Eds.), Handbook of clinical neuropsychology (Vol. 1, pp. 418–452). New York: Wiley-Interscience.

Cohn, R. (1968). Developmental dyscalculia. Pediatric Clinics of North America, 15, 651–668.

Dahmen, W., Hartje, W., Büssing, A., & Sturm, W. (1982). Disorders of calculation in aphasic patients—spatial and verbal components. Neuropsychologia, 20, 145–153.

Davis, A. E., & Wada, J. A. (1977). Hemispheric asymmetries in human infants: Spectral analysis of flash and click evoked potentials. Brain and Language, 4, 23–31.

DeLuca, J. W., Rourke, B. P., & Del Dotto, J. E. (1991). Subtypes of arithmetic-disabled children: Cognitive and personality dimensions. In B. P. Rourke (Ed.), Neuropsychological validation of learning disability subtypes (pp. 180–219). New York: Guilford.

Denckla, M. B. (1973). Research needs in learning disabilities: A neurologist's point of view. Journal of Learning Disabilities, 6, 44–50.

De Renzi, E. (1978). Hemispheric asymmetry as evidenced by spatial disorders. In M. Kinsbourne (Ed.), Asymmetrical function of the brain (pp. 49–85). Cambridge, England: Cambridge University Press.

De Renzi, E., & Faglioni, P. (1967). The relationship between visuo-spatial impairment and constructional apraxia. Cortex, 3, 327–342.

Dool, C. B., Stelmack, R. M., & Rourke, B. P. (1993). Event-related potentials in children with learning disabilities. Journal of Clinical Child Psychology, 22, 387–398.

Doyle, J. C., Ornstein, R., & Galin, D. (1974). Lateral specialization of cognitive mode: II. EEG frequency analysis. Psychophysiology, 11, 567–578.

Earle, J. B. (1985). The effects of arithmetic task difficulty and performance level on EEG alpha asymmetry. Neuropsychologia, 23, 233–242.

Ewing-Cobbs, L., Fletcher, J. M., & Levin, H. S. (1995). Traumatic brain injury. In B. P. Rourke (Ed.), Syndrome of nonverbal learning disabilities: Manifestations in neurological disease, disorder, and dysfunction (pp. 433–459). New York: Guilford.

Fisk, J. L., & Rourke, B. P. (1979). Identification of subtypes of learning-disabled children at three age levels: A neuropsychological, multivariate approach. Journal of Clinical Neuropsychology, 1, 289–310.

Fletcher, J. M., Brookshire, B. L., Bohan, T. P., Brandt, M., & Davidson, K. (1995). Hydrocephalus. In B. P. Rourke (Ed.), Syndrome of nonverbal learning disabilities: Neurodevelopmental manifestations (pp. 206–238). New York: Guilford.

Gaddes, W. H. (1985). Learning disabilities and brain function: A neuropsychological approach (2nd ed.). New York: Springer-Verlag.

Galaburda, A. M., LeMay, M., Kemper, T. L., & Geschwind, N. (1978). Right–left asymmetries in the brain. Science, 199, 852–856.

Galin, D., & Ellis, R. (1975). Asymmetry in evoked potentials as an index of lateralized cognitive processes: Relation to EEG alpha asymmetry. Neuropsychologia, 13, 45–50.

Geary, D. C. (1993). Mathematical disabilities: Cognitive, neuropsychological, and genetic components. Psychological Bulletin, 114, 345–362.

Gerstmann, J. (1940). Syndrome of finger agnosia, disorientation for right and left, agraphia, and acalculia. Archives of Neurology and Psychiatry, 44, 398–408.

Geschwind, N., & Levitsky, W. (1968). Human brain left–right asymmetries in temporal speech regions. Science, 161, 186–187.

Goldberg, E., & Costa, L. D. (1981). Hemisphere differences in the acquisition and use of descriptive systems. Brain and Language, 14, 144–173.

Grafman, J., Passafiume, D., Faglioni, P., & Boller, F. (1982). Calculation disturbances in adults with focal hemispheric damage. Cortex, 18, 37–50.

Grewel, F. (1952). Acalculia. Brain, 75, 397–407.

Grigsby, J. P., Kemper, M. B., & Hagerman, R. J. (1987). Developmental Gerstmann syndrome without aphasia in Fragile X syndrome. Neuropsychologia, 25, 881–891.

Hartje, W. (1987). The effect of spatial disorders on arithmetical skills. In G. Deloche & X. Seron (Eds.), Mathematical disabilities: A cognitive neuropsychological perspective (pp. 121–135). Hillsdale, NJ: Erlbaum.

Hécaen, H. (1962). Clinical symptomatology in right and left hemispheric lesions. In V. B. Mountcastle (Ed.), Interhemispheric relations and cerebral dominance (pp. 215–243). Baltimore: Johns Hopkins.

Hécaen, H., & Angelergues, R. (1963). La cécité psychique [Psychic blindness]. Paris: Masson et Cie.

Hécaen, H., Angelergues, R., & Houillier, S. (1961). Les variétiés cliniques des acalculies au cours des lésions rétrorolandiques: Approche statistique du problème [The clinical varieties of the acalculies in retrorolandic lesions: A statistical approach to the problem]. Revue Neurologique, 105, 85–103.

Heimburger, R. F., DeMyer, W. C., & Reitan, R. M. (1964). Implications of Gerstmann's syndrome. Journal of Neurology, Neurosurgery, and Psychiatry, 27, 52–57.

Henschen, S. E. (1919). Clinical and anatomical contributions on brain pathology. Archives of Neurology and Psychiatry, 13, 226–249.

Jastak, J. F., & Jastak, S. R. (1965). The Wide Range Achievement Test. Wilmington, DE: Guidance Associates.

Keller, C. E., & Sutton, J. P. (1991). Specific mathematics disorders. In J. E. Obrzut & G. W. Hynd (Eds.), Neuropsychological foundations of learning disabilities: A *handbook of issues, methods, and practice (pp. 549-571). San Diego: Academic Press.*

Kertesz, A., & Dobrowolski, S. (1981). Right-hemisphere deficits, lesion size and location. *Journal of Clinical Neuropsychology, 3, 283–299.*

Kimura, D. (1963). Speech lateralization in young children as determined by an auditory test. *Journal of Comparative and Physiological Psychology, 56, 899–902.*

Kinsbourne, M. (1968). Developmental Gerstmann syndrome. *Pediatric Clinics of North America, 15, 771–778.*

Kinsbourne, M., & Warrington, E. K. (1963). The developmental Gerstmann syndrome. *Archives of Neurology, 8, 490–501.*

Kolb, B., & Whishaw, I. Q. (1990). *Fundamentals of human neuropsychology (3rd ed.).* New York: W. H. Freeman.

Kosc, L. (1974). Developmental dyscalculia. *Journal of Learning Disabilities, 7, 165–177.*

Levin, H. S., Goldstein, F. C., & Spiers, P. A. (1993). Acalculia. In K. M. Heilman & E. Valenstein (Eds.), *Clinical neuropsychology (3rd ed., pp. 91–122).* New York: Oxford University Press.

Lezak, M. D. (1983). *Neuropsychological assessment (2nd ed.).* New York: Oxford University Press.

Licht, R., Bakker, D. J., Kok, A., & Bouma, A. (1992). Grade-related changes in event-related potentials (ERPs) in primary school children: Differences between two reading tasks. *Journal of Clinical and Experimental Neuropsychology, 14, 193–210.*

PeBenito, R., Fisch, C. B., & Fisch, M. L. (1988). Developmental Gerstmann's syndrome. *Archives of Neurology, 45, 977–982.*

Piaget, J. (1954). *The construction of reality in the child.* New York: Basic Books.

Rebert, C. S., Wexler, B. N., & Sproul, A. (1978). EEG asymmetry in educationally handicapped children. *Electroencephalography & Clinical Neurophysiology, 45, 436–442.*

Reitan, R. M., & Davison, L. A. (Eds.). (1974). *Clinical neuropsychology: Current status and applications.* New York: Wiley.

Reuter-Lorenz, P. A., Kinsbourne, M., & Moscovitch, M. (1990). Hemispheric control of spatial attention. *Brain and Cognition, 12, 240–266.*

Rourke, B. P. (1982). Central processing deficiencies in children: Toward a developmental neuropsychological model. *Journal of Clinical Neuropsychology, 4, 1–18.*

Rourke, B. P. (1987). Syndrome of nonverbal learning disabilities: The final common pathway of white-matter disease/dysfunction? *The Clinical Neuropsychologist, 1, 209–234.*

Rourke, B. P. (1988). The syndrome of nonverbal learning disabilities: Developmental manifestations in neurological disease, disorder, and dysfunction. *The Clinical Neuropsychologist, 2, 293–330.*

Rourke, B. P. (1989). *Nonverbal learning disabilities: The syndrome and the model.* New York: Guilford.

Rourke, B. P. (1993). Arithmetic disabilities, specific and otherwise: A neuropsychological perspective. *Journal of Learning Disabilities, 26, 214–226.*

Rourke, B. P. (Ed.). (1995). *Syndrome of nonverbal learning disabilities: Neurodevelopmental manifestations.* New York: Guilford.

Rourke, B. P., Bakker, D. J., Fisk, J. L., & Strang, J. D. (1983). *Child neuropsychology: An introduction to theory, research, and clinical practice.* New York: Guilford.

Rourke, B. P., Dietrich, D. M., & Young, G. C. (1973). Significance of WISC Verbal–Performance discrepancies for younger children with learning disabilities. *Perceptual and Motor Skills, 36,* 275–282.

Rourke, B. P., & Finlayson, M. A. (1978). Neuropsychological significance of variations in patterns of academic performance: Verbal and visual–spatial abilities. *Journal of Abnormal Child Psychology, 6,* 121–133.

Rourke, B. P., & Fisk, J. L. (1988). Subtypes of learning-disabled children: Implications for a neurodevelopmental model of differential hemispheric processing. In D. L. Molfese & S. J. Segalowitz (Eds.), *Brain lateralization in children: Developmental implications* (pp. 547–565). New York: Guilford.

Rourke, B. P., & Fuerst, D. R. (1995). Cognitive processing, academic achievement, and psychosocial functioning: A neuropsychological perspective. In D. Cicchetti & D. Cohen (Eds.), *Manual of developmental psychopathology* (Vol. 1, pp. 391–423). New York: Wiley.

Rourke, B. P., & Strang, J. D. (1978). Neuropsychological significance of variations in patterns of academic performance: Motor, psychomotor, and tactile–perceptual abilities. *Journal of Pediatric Psychology, 3,* 62–66.

Rourke, B. P., & Telegdy, G. A. (1971). Lateralizing significance of WISC Verbal–Performance discrepancies for older children with learning disabilities. *Perceptual and Motor Skills, 33,* 875–883.

Rourke, B. P., & Tsatsanis, K. M. (1995). Memory disturbances of children with learning disabilities: A neuropsychological analysis of academic achievement subtypes. In A. Baddeley, B. Wilson, & F. Watts (Eds.), *Handbook of memory disorders* (pp. 503–533). New York: Wiley.

Rourke, B. P., Young, G. C., & Flewelling, R. W. (1971). The relationships between WISC Verbal–Performance discrepancies and selected verbal, auditory–perceptual, visual–perceptual, and problem-solving abilities in children with learning disabilities. *Journal of Clinical Psychology, 27,* 475–479.

Saxe, G. B., & Shaheen, S. (1981). Piagetian theory and the atypical case: An analysis of the developmental Gerstmann syndrome. *Journal of Learning Disabilities, 14,* 131–135, 172.

Semmes, J. (1968). Hemispheric specialization: A possible clue to mechanism. *Neuropsychologia, 6,* 11–26.

Semrud-Clikeman, M., & Hynd, G. W. (1990). Right hemisphere dysfunction in nonverbal learning disabilities: Social, academic, and adaptive functioning in adults and children. *Psychological Bulletin, 107,* 196–209.

Share, D. L., Moffitt, T. E., & Silva, P. A. (1988). Factors associated with arithmetic-and-reading disability and specific arithmetic disability. *Journal of Learning Disabilities, 21,* 313–320.

Singer, H., & Low, A. (1933). Acalculia. *Archives of Neurology and Psychiatry, 29,* 467–498.

Spellacy, F., & Peter, B. (1978). Dyscalculia and elements of the developmental Gerstmann syndrome in school children. *Cortex, 14,* 197–206.

Spiers, P. A. (1987). Acalculia revisited: Current issues. In G. Deloche & X. Seron (Eds.), *Mathematical disabilities: A cognitive neuropsychological perspective* (pp. 1–25). Hillsdale, NJ: Erlbaum.

Strang, J. D., & Rourke, B. P. (1983). Concept-formation/non-verbal reasoning abilities of children who exhibit specific academic problems with arithmetic. *Journal of Clinical Child Psychology, 12,* 33–39.

Wechsler, D. (1981). *Wechsler Adult Intelligence Scale–Revised (manual).* San Antonio, TX: Psychological Corp.

White, J. L., Moffitt, T. E., & Silva, P. A. (1992). Neuropsychological and socio-emotional correlates of specific-arithmetic disability. *Archives of Clinical Neuropsychology, 7,* 1–16.

CHAPTER 4

Ackerman, P. T., Anhalt, J. M., & Dykman, R. A. (1986). Arithmetic automatization failure in children with attention and reading disorders: Associations and sequilae. *Journal of Learning Disabilities, 19,* 222–232.

Baroody, A. J., & Hume, J. (1991). Meaningful mathematics instruction: The case of fractions. *Remedial and Special Education, 12(3),* 54–68.

Bartel, N. R. (1982). Problems in mathematics achievement. In D. D. Hammill & N. R. Bartel (Eds.), *Teaching children with learning and behavior problems* (3rd ed., pp. 173–198). Boston: Allyn & Bacon.

Bateman, B. (1992). Learning disabilities: The changing landscape. *Journal of Learning Disabilities, 25,* 29–36.

Bos, C. S., & Vaughn, S. (1994). *Strategies for teaching students with learning and behavior problems.* Boston: Allyn & Bacon.

Brownell, M. T., Mellard, D. F., & Deshler, D. D. (1993). Differences in the learning and transfer performances between students with learning disabilities and other low-achieving students on problem-solving tasks. *Learning Disability Quarterly, 16,* 138–156.

Carnine, D. (1991). Curricular interventions for teaching higher order thinking to all students: Introduction to the special series. *Journal of Learning Disabilities, 24,* 261–269.

Carnine, D. (1992). The missing link in improving schools—reforming educational leaders. *Direct Instruction News, 11(3),* 25–35.

Carnine, D. (1994). Ideologies, practices, and their implications for special education. *The Journal of Special Education, 28,* 356–367.

Carnine, D. W., & Kameenui, E. J. (1990). The General Education Initiative and children with special needs: A false dilemma in the face of true problems. *Journal of Learning Disabilities, 23,* 141–144, 148.

Cawley, J. F., Baker-Kroczynski, S., & Urban, A. (1992). Seeking excellence in mathematics education for students with mild disabilities. Teaching Exceptional Children, 24, 40–43.

Cawley, J. F., Fitzmaurice-Hayes, A., & Shaw, R. (1988). Mathematics for the mildly handicapped—A guide to curriculum and instruction. Boston: Allyn & Bacon.

Cawley, J. F., & Miller, J. H. (1989). Cross-sectional comparisons of the mathematical performance of children with learning disabilities: Are we on the right track toward comprehensive programming? Journal of Learning Disabilities, 23, 250–254, 259.

Cawley, J. F., Miller, J. H., & School, B. A. (1987). A brief inquiry of arithmetic word-problem solving among learning disabled secondary students. Learning Disabilities Focus, 2, 87–93.

Cawley, J. F., & Parmar, R. S. (1990). Issues in mathematics curriculum for handicapped students. Academic Therapy, 25, 507–521.

Chambers, D. L. (1994). The right algebra for all. Educational Leadership, 51(6), 85–86.

Cherkes-Julkowski, M. (1985). Information processing: A cognitive view. In J. Cawley (Ed.), Cognitive strategies and mathematics for the learning disabled (pp. 117–138). Austin, TX: PRO-ED.

Cohen, S. B., Safran, J., & Polloway, E. (1980). Minimum competency testing: Implications for mildly retarded students. Education and Training of the Mentally Retarded, 15, 250–255.

Conte, R. (1991). Attention disorders. In B. Wong (Ed.), Learning about learning disabilities (pp. 60–103). San Diego: Academic Press.

Dixon, B. (1994). Research guidelines for selecting mathematics curriculum. Effective School Practices, 13(2), 47–55.

Dossey, J. A., Mullis, I. V. S., Lindquist, M. M., & Chambers, D. L. (1988). The mathematics report card: Are we measuring up? Princeton, NJ: Educational Testing Service. (ERIC Document Reproduction Service No. ED 300 206)

Engelmann, S., Carnine, D., & Steely, D. G. (1991). Making connections in mathematics. Journal of Learning Disabilities, 24, 292–303.

Englert, C. S., Culatta, B. E., & Horn, D. G. (1987). Influence of irrelevant information in addition word problems on problem solving. Learning Disability Quarterly, 10, 29–36.

Fleischner, J. E., Garnett, K., & Shepherd, M. (1982). Proficiency in arithmetic basic fact computation by learning disabled and nondisabled children. Focus on Learning Problems in Mathematics, 4, 47–55.

Fuchs, D., & Fuchs, L. (1988). An evaluation of the Adaptive Learning Environments Model. Exceptional Children, 55, 115–127.

Fuchs, D., & Fuchs, L. (1994). Inclusive schools movement and the radicalization of special education reform. Exceptional Children, 60, 294–309.

Gamoran, A., & Berends, M. (1987). The efforts of stratification in secondary schools. Review of Educational Research, 57, 415–435.

Garnett, K. (1992). Developing fluency with basic number facts: Intervention for students with learning disabilities. Learning Disabilities Research & Practice, 7, 210–216.

Gartner, A., & Lipsky, D. K. (1987). Beyond special education: Toward a quality system for all students. *Harvard Educational Review, 57,* 367–395.

Greenstein, J., & Strains, P. S. (1977). The utility of the Keymath Diagnostic Arithmetic Test for adolescent learning disabled students. *Psychology in the Schools, 14,* 275–282.

Goldman, S. R. (1989). Strategy instruction in mathematics. *Learning Disability Quarterly, 12,* 43–55.

Hallahan, D. P., Keller, C. E., McKinney, J. D., Lloyd, J. W., & Bryan, T. (1988). Examining the research base of the Regular Education Initiative: Efficacy studies and the ALEM. *Journal of Learning Disabilities, 21,* 29–35.

Hofmeister, A. M. (1993). Elitism and reform in school mathematics. *Remedial and Special Education, 14(6),* 8–13.

Howell, K. W., Fox, S. L., & Morehead, M. K. (1993). *Curriculum-based evaluation: Teaching and decision making* (2nd ed.). Pacific Grove, CA: Brooks/Cole.

Hutchinson, N. L. (1993). Students with disabilities and mathematics education reform—Let the dialogue begin. *Remedial and Special Education, 14(6),* 20–23.

Kavale, K. A., Fuchs, D., & Scruggs, T. E. (1994). Setting the record straight on learning disability and low achievement: Implications for policy making. *Learning Disabilities Research & Practice, 9,* 70–77.

Kelly, B., Gersten, R., & Carnine, D. (1990). Student error patterns as a function of curriculum design: Teaching fractions to remedial high school students and high school students with learning disabilities. *Journal of Learning Disabilities, 23,* 23–29.

Kulak, A. G. (1993). Parallels between math and reading disability: Common issues and approaches. *Journal of Learning Disabilities, 26,* 666–673.

Lapointe, A. E., Mead, N. A., & Phillips, G. W. (1989). *A world of differences: An international assessment of mathematics and science.* Princeton, NJ: Educational Testing Service. (ERIC Document Reproduction Service No. ED 309 068)

Lee, W. M., & Hudson, F. G. (1981). *A comparison of verbal problem-solving in arithmetic of learning disabled and non-learning disabled seventh grade males* (Research Report No. 43). Lawrence: Institute for Research in Learning Disabilities, University of Kansas.

Lerner, J. (1993). *Learning disabilities* (6th ed.). Boston: Houghton Mifflin.

Lichtenstein, S. (1993). Transition from school to adulthood: Case studies of adults with learning disabilities who dropped out of school. *Exceptional Children, 59,* 336–347.

Lovitt, T. C. (1989). *Introduction to learning disabilities.* Boston: Allyn & Bacon.

Masters, L. F., Mori, B. A., & Mori, A. A. (1993). *Teaching secondary students with mild learning and behavior problems.* Austin, TX: PRO-ED.

Mastropieri, M. A., Scruggs, T. E., & Shiah, S. (1991). Mathematics instruction for learning disabled students: A review of research. *Learning Disabilities Research & Practice, 6,* 89–98.

Mather, N., & Roberts, R. (1994). Learning disabilities: A field in danger of extinction? *Learning Disabilities Research & Practice, 9,* 49–58.

McLeod, T., & Armstrong, S. (1982). Learning disabilities in mathematics-skill deficits and remedial approaches at the intermediate and secondary grades. Learning Disability Quarterly, 5, 305–311.

Mercer, C. D. (1992). Students with learning disabilities (4th ed.). New York: Macmillan.

Mercer, C. D., Harris, C. A., & Miller, S. P. (1993). Reforming reforms in mathematics. Remedial and Special Education, 14(6), 14–19.

Mercer, C. D., Jordan, L., & Miller, S. P. (1994). Implications of constructivism for teaching math to students with moderate to mild disabilities. The Journal of Special Education, 28, 290–306.

Mercer, C. D., & Miller, S. P. (1992). Teaching students with learning problems in math to acquire, understand, and apply basic math facts. Remedial and Special Education, 13(3), 19–35, 61.

Montague, M., & Applegate, B. (1993). Middle school students' mathematical problem solving: An analysis of think-aloud protocols. Learning Disability Quarterly, 16, 19–30.

Montague, M., Bos, C., & Doucette, M. (1991). Affective, cognitive, and metacognitive attributes of eighth-grade mathematical problem solvers. Learning Disabilities Research & Practice, 6, 145–151.

National Council of Supervisors of Mathematics. (1988). Twelve components of essential mathematics. Minneapolis: Author.

National Council of Teachers of Mathematics. (1989). Curriculum and evaluation standards for school mathematics. Reston, VA: Author.

Parmar, R. S., & Cawley, J. F. (1991). Challenging the routines and passivity that characterize arithmetic instruction for children with mild handicaps. Remedial and Special Education, 12(5), 23–32, 43.

Parmar, R. S., Cawley, J. F., & Miller, J. H. (1994). Differences in mathematics performance between students with learning disabilities and students with mild retardation. Exceptional Children, 60, 549–563.

Resnick, L., & Resnick, D. (1985). Standards, curriculum and performance: A historical perspective. Educational Researcher, 14, 5–20.

Rivera, D. M. (1993). Examining mathematics reform and the implications for students with mathematics disabilities. Remedial and Special Education, 14(6), 24–27.

Roberts, C., & Zubrick, S. (1992). Factors influencing the social status of children with mild academic disabilities in regular classrooms. Exceptional Children, 59, 192–202.

Romberg, T. A. (1993). NCTM's Standards: A rallying flag for mathematics teachers. Educational Leadership, 50(5), 36–41.

Scheid, K. (1990). Cognitive-based methods for teaching mathematics to students with learning problems. Columbus, OH: LINC Resources.

Schumaker, J. B., & Deshler, D. D. (1988). Implementing the Regular Education Initiative in secondary schools: A different ball game. Journal of Learning Disabilities, 21, 36–42.

Semmel, M. I., Abernathy, T. V., Butera, G., & Lesar, S. (1991). Teacher perceptions of the Regular Education Initiative. Exceptional Children, 58, 9–24.

Silbert, J., & Carnine, D. (1990). The mathematics curriculum—Standards, textbooks, and pedagogy: A case study of fifth grade division. ADI News, 10(1), 39–47.

Silver, L. B. (1991). The Regular Education Initiative: A déjà vu remembered with sadness and concern. Journal of Learning Disabilities, 24, 389–390.

Simmons, D. C., Fuchs, D., & Fuchs, L. S. (1991). Instructional and curricular requisites of mainstreamed students with learning disabilities. Journal of Learning Disabilities, 24, 354–359, 353.

Slavin, R. (1991). Educational psychology. Englewood Cliffs, NJ: Prentice-Hall.

Slavin, R. E., & Madden, N. A. (1989). What works for students at risk: A research synthesis. Educational Leadership, 46(5), 4–13.

Smith, C. R. (1994). Learning disabilities: The interaction of learner, task, and setting (3rd ed.). Boston: Allyn & Bacon.

Snell, M. E. (1991). Schools are for all kids: The importance of integration for students with severe disabilities and their peers. In J. W. Lloyd, A. C. Repp, & N. N. Singh (Eds.), The Regular Education Initiative: Alternative perspectives on concepts, issues, and models (pp. 133–148). Sycamore, IL: Sycamore.

Sovchik, R. J. (1989). Teaching mathematics to children. New York: Harper & Row.

Sprick, R. S. (1987). Solutions to elementary discipline problems [Audiotapes]. Eugene, OR: Teaching Strategies.

Strang, J. D., & Rourke, B. P. (1985). Arithmetic disability subtypes: The neuropsychological significance of specific arithmetical impairment in childhood. In B. P. Rourke (Ed.), Neuropsychology of learning disabilities: Essentials of subtype analysis (pp. 167–183). New York: Guilford.

Swanson, H. L. (1990). Instruction derived from the strategy deficit model: Overview of principles and procedures. In T. Scruggs & B. Wong (Eds.), Intervention research in learning disabilities (pp. 34–65). New York: Springer-Verlag.

Torgesen, J. (1990). Studies of children with learning disabilities who perform poorly on memory span tasks. In J. K. Torgesen (Ed.), Cognitive and behavioral characteristics of children with learning disabilities (pp. 41–58). Austin, TX: PRO-ED.

Tyson, H., & Woodward, A. (1989). Why students aren't learning very much from textbooks. Educational Leadership, 47(3), 14–17.

Usiskin, Z. (1993). Lessons from the Chicago mathematics project. Educational Leadership, 50(8), 14–18.

Wagner, M. (1990, April). The school programs and school performance of secondary students classified as learning disabled: Findings from the National Longitudinal Transition Study of Special Education Students. Paper presented at the meetings of Division G, American Educational Research Association, Boston.

Warner, M., Alley, G., Schumaker, J., Deshler, D., & Clark, F. (1980). An epidemiological study of learning disabled adolescents in secondary schools: Achievement and ability, socioeconomic status and school experiences (Report No. 13). Lawrence: University of Kansas Institute for Research in Learning Disabilities.

Wilson, C. L., & Sindelar, P. T. (1991). Direct instruction in math word problems: Students with learning disabilities. Exceptional Children, 57, 512–519.

Woodward, J. (1991). Procedural knowledge in mathematics: The role of the curriculum. Journal of Learning Disabilities, 24, 242–251.

Zawaiza, T. R. W., & Gerber, M. M. (1993). Effects of explicit instruction on math word-problem solving by community college students with learning disabilities. Learning Disability Quarterly, 16, 64–79.

Zentall, S. S., & Ferkis, M. A. (1993). Mathematical problem solving for children with ADHD, with and without learning disabilities. Learning Disability Quarterly, 16, 6–18.

Zentall, S. S., & Zentall, T. R. (1983). Optimal stimulation: A model of disordered activity and performance in normal and deviant children. Psychological Bulletin, 94, 446–471.

CHAPTER 5

American Educational Research Association & National Council on Measurement in Education (1955). Technical recommendations for achievement tests. Washington, DC: National Education Association.

American Psychological Association. (1954). Technical recommendations for psychological tests and diagnostic techniques. Washington, DC: Author.

American Psychological Association. (1966). Standards for educational and psychological tests and manuals. Washington, DC: Author.

Baroody, A., & Ginsburg, H. (1991). Mathematics assessment. In H. L. Swanson (Ed.), Handbook on the assessment of learning disabilities (pp. 177–227). Austin, TX: PRO-ED.

Binet, A., & Simon, T. (1905). Methods nouvelles pour le diagnostic de niveau intellectuel des anormaux. Annec Psychologique, 11, 191–244.

Bryant, B. R., & Maddox, T. (in press). Using alternative assessment techniques to plan for and evaluate mathematics instruction. LD Forum.

Bryant, B. R., Patton, J., & Dunn, C. (1991). Scholastic abilities test for adults. Austin, TX: PRO-ED.

Bryant, B. R., Taylor, R., & Rivera, D. P. (1994, October). Mathematics assessment and the NCTM standards: Implications for students with learning disabilities. Paper presented at the International Conference on Learning Disabilities, San Diego.

Buros, O. K. (Ed.). (1975). Mathematics tests and reviews. Highland Park, NJ: Gryphon.

Burt, C. (1922). Mental and scholastic tests. London: P. S. King and Son.

Buswell, G. T., & John, L. (1926). Diagnostic studies in arithmetic. (Supplementary Educational Monograph No. 30). Chicago: The University of Chicago.

Case, L. P., Harris, K. R., & Graham, S. (1992). Improving the mathematical skills of students with learning disabilities: Self-regulated strategy development. The Journal of Special Education, 26, 1–19.

Connolly, A. J. (1988). KeyMath–Revised. Circle Pines, MN: American Guidance Service.

Connolly, A., Nactman, W., & Prichett, E.M. (1971). KeyMath diagnostic arithmetic test. Circle Pines, MN: American Guidance Service.

Cooper, B. (1985). *Renegotiating secondary school mathematics: A study of curriculum change and stability.* London: Falmer.

Coutinho, M., & Malouf, D. (1993). Performance assessment and children with disabilities: Issues and possibilities. *Teaching Exceptional Children, 25(4),* 62–67.

Cronin, M. E. (1985). Assessment techniques and practices for classroom behaviors, social/emotional factors, and attitudes. In J.F. Cawley (Ed.), *Practical mathematics: Appraisal of the learning disabled* (pp. 215–247). Rockville, MD: Aspen.

Deno, S. L. (1985). Curriculum-based measurement: The emerging alternative. *Exceptional Children, 52,* 219–232.

DeRuiter, J. A., & Wansart, W. L. (1982). *Psychology of learning disabilities.* Rockville, MD: Aspen.

Fuchs, L., & Fuchs, D. (1988). Curriculum-based measurement: A methodology for evaluating and improving student programs. *Diagnostique, 14,* 3–13.

Fuchs, L., Fuchs, D., Hamlett, C., & Allinder, R. (1989). The reliability and validity of skills analysis within curriculum-based measurement. *Diagnostique, 14,* 203–221.

Fuchs, L., Fuchs, D., Hamlett, C., & Allinder, R. (1991). The contribution of skills analysis to curriculum-based measurement in spelling. *Exceptional Children, 57,* 441–452.

Fuchs, L., Fuchs, D., Hamlett, C. L., Thompson, A., Roberts, P. H., Kubek, P., & Stecker, P. (1994). Technical features of a mathematics concepts and applications curriculum-based measurement system. *Diagnostique, 19(4),* 23–50.

Fuchs, L. S., Hamlett, C., & Fuchs, D. (1990). *Monitoring basic skills progress: Basic math.* Austin, TX: PRO-ED.

Gilliland, A. R., & Jordan, R. H. (1924). *Educational measurements and the classroom teacher.* New York: Century.

Ginsburg, H. P. (1987). The development of arithmetic thinking. In D. D. Hammill (Ed.), *Assessing the abilities and instructional needs of students* (pp. 423–440). Austin, TX: PRO-ED.

Ginsburg, H. P., & Baroody, A. (1990). *Test of early mathematics abilities* (2nd ed.). Austin, TX: PRO-ED.

Ginsburg, H., & Mathews, S. C. (1984). *Diagnostic test of arithmetic strategies.* Austin, TX: PRO-ED.

Goodlad, J. I., Stoephasius, R. V., & Klein, M. F. (1966). *The changing school curriculum.* New York: Fund for the Advancement of Education.

Greene, H. A., Jorgensen, A. L., & Gerberich, J. R. (1953). *Measurement and evaluation in the elementary school.* New York: Longsman, Green.

Hammill, D., Ammer, J., Cronin, M., Mandlebaum, L., & Quinby, S. (1987). *Quick-score achievement test.* Austin, TX: PRO-ED.

Hammill, D. D., Brown, L., & Bryant, B. R. (1992). *A consumer's guide to tests in print* (2nd ed.). Austin, TX: PRO-ED.

Hammill, D. D., & Bryant, B. R. (1991). Standardized assessment and academic intervention. In H. L. Swanson (Ed.), *Handbook on the assessment of learning disabilities: Theory, research, and practice* (pp. 373–406). Austin, TX: PRO-ED.

Hammill, D. D., & Larsen, S. (1974). The effectiveness of psycholinguistic training. *Exceptional Children, 41,* 5–15.

Hasselbring, T. S., Goin, L. I., & Bransford, J. D. (1987). Effective mathematics instruction: Developing automaticity. Teaching Exceptional Children, 19(3), 30–33.

Hofmeister, A. M. (1993). Elitism and reform in school mathematics. Remedial and Special Education, 14(6), 8–13.

Hutchinson, N. L. (1993). Second invited response: Students with disabilities and mathematics education reform—Let the dialogue begin. Remedial and Special Education, 14(6), 20–23.

Jenkins, J. R., & Jewell, M. (1992). An examination of the concurrent validity of the Basic Academic Skills Samples (BASS). Diagnostique, 17(4), 273–288.

Johnson, D. J., & Mykelbust, H. R. (1967). Learning disabilities: Educational principles and practices. New York: Grune & Stratton.

Kaufman, A. S., & Kaufman, N. L. (1985). Kaufman test of educational achievement–Comprehensive form. Circle Pines, MN: American Guidance Service.

Kennedy, L. M., & Tipps, S. (1994). Guiding children's learning of mathematics (7th ed.). Belmont, CA: Wadsworth.

Kirk, S. A., McCarthy, J. J., & Kirk, W. D. (1968). Illinois test of psycholinguistic abilities. Champaign: University of Illinois Press.

Kliebard, H. M. (1995). The struggle for the American curriculum: 1893-1958 (2nd ed.). New York: Herbert M. Kliebard.

Lopez-Reyna, N. A., Bay, M., & Patrikakou, E. N. (1996). Use of assessment procedures: Learning disabilities teachers' perspectives. Diagnostique, 21(2), 35–49.

Markwardt, F. C. (1989). Peabody individual achievement test–Revised. Circle Pines, MN: American Guidance Service.

McKinney, J. D., Osbourne, S. S., & Schulte, A. C. (1993). Academic consequences of learning disability: Longitudinal prediction of outcomes at 11 years of age. Learning Disability Research & Practice, 8(1), 19–27.

McNeil, J. D. (1977). Curriculum: A comprehensive introduction. Boston: Little, Brown.

Mercer, C. D., Harris, C. A., & Miller, S. P. (1993). First invited response: Reforming reforms in mathematics. Remedial and Special Education, 14(6), 14–19.

Mercer, C. D., & Mercer, A. R. (1993). Teaching students with learning problems (4th ed.). New York: Merrill-Macmillan.

Miller, S. P., & Mercer, C. D. (1993). Using data to learn about concrete–semiconcrete–abstract instructions for students with math disabilities. Learning Disability Research & Practice, 8(2), 89–96.

Montague, M., & Applegate, B. (1993). Mathematical problem solving characteristics of middle school students with learning disabilities. The Journal of Special Education, 27, 175–201.

Montague, M., Applegate, B., & Marquard, K. (1993). Cognitive strategy instruction and mathematical problem-solving performance of students with learning disabilities. Learning Disability Research & Practice, 8(4), 223–232.

National Council of Teachers of Mathematics. (1989). Curriculum and evaluation standards for school mathematics. Reston, VA: Author.

Newcomer, P. (1990) Diagnostic achievement battery. Austin, TX: PRO-ED.

Newcomer, P., & Bryant, B. R. (1992). Diagnostic achievement test for adolescents (2nd ed.). Austin, TX: PRO-ED.

Parmar, R. (1992). Protocol analysis of strategies used by students with mild disabilities when solving arithmetic word problems. *Diagnostique, 17(4)*, 227–243.

Parmar, R. S., Frazita, R., & Cawley, J. F. (1996). Mathematics assessment for students with mild disabilities: An exploration of content validity. *Learning Disability Quarterly, 19(2)*, 127–136.

Paulson, F. L, Paulson, P. R., & Meyer, C. A. (1991). What makes a portfolio a portfolio? *Educational Leadership, 48(5)*, 60–63.

Psychological Corporation (1983). *Basic achievement skills individual screener*. San Antonio, TX: Author.

Ragan, W. B., & Shepherd, G. D. (1971). *Modern elementary curriculum* (4th ed.). New York: Holt, Rinehart & Winston.

Resnick, L. B., & Ford, W. W. (1981). *The psychology of mathematics for instruction*. Hillsdale, NJ: Erlbaum.

Rivera, D. (1993). Third invited response: Examining mathematics reform and the implications for students with mathematics disabilities. *Remedial and Special Education, 14(6)*, 24–27.

Rivera, D. (1994). Portfolio assessment. *LD Forum, 19(4)*, 14–17.

Rivera, D. P., & Bryant, B. R. (1992). Mathematics instruction for students with special needs. *Intervention in School and Clinic, 28*, 71–86.

Rivera, D. P., & Smith, D. D. (1996). *Teaching students with learning and behavior problems* (3rd ed.). Boston: Allyn & Bacon.

Rivera, D. P., Taylor, R. L., & Bryant, B. R. (1994–1995). Review of current trends in mathematics assessment for students with mild disabilities. *Diagnostique, 20(1–4)*, 143–174.

Ruch, G. M., Knight, F. B., Greene, H. A., & Studebaker, J. W. (1925). *Compass diagnostic tests in arithmetic*. Chicago: Scott, Foresman.

Salvia, J., & Ysseldyke, J. E. (1995). *Assessment* (6th ed.). Boston: Houghton Mifflin.

Sandler, A. D., Hooper, S. R., Scarborough, A. A., Watson, T. E., & Levine, M. D. (1993). Adolescents talking about thinking: Preliminary findings of a self-report instrument for the assessment of cognition and learning. *Diagnostique, 19(1)*, 361–376.

Shinn, M. R., & Hubbard, D. D. (1993). Curriculum-based measurement and problem-solving assessment: Basic procedures and outcomes. In E. L. Meyen, G. A. Vergason, & R. J. Whelan (Eds.), *Challenges facing special education* (pp. 193–226). Denver: Love.

Slate, J. R., Jones, C. H., Graham, L. P., & Bower, J. (1994). Correlations of WISC-III, WRAT, KM-R, PPVT-R scores in students with specific learning disabilities. *Learning Disability Research & Practice, 9(2)*, 104–107.

Swicegood, P. (1994). Portfolio-based assessment practices. *Intervention in School and Clinic, 30(1)*, 6–15.

Taylor, R. L. (1993). *Assessment of exceptional children* (3rd ed.). Needham Heights, MA: Allyn & Bacon.

Taylor, R. L., Tindal, G., Fuchs, L., & Bryant, B. R. (1993). Assessment in the nineties: A possible glance into the future. *Diagnostique, 18(2)*, 113–122.

Thorndike, E. L. (1912). *Education: A first book*. New York: Macmillan.

Thorndike, E. L. (1929). *Elementary principles of education*. New York: Macmillan.

Thorndike, R. L., & Hagen, E. (1955). *Measurement and evaluation in psychology and education*. New York: Wiley.

Tiegs, E. W. (1931). *Tests and measurement for teachers*. Cambridge, MA: Riverside.

Tucker, J. (1985). Curriculum-based assessment: An introduction. *Exceptional Children, 52*, 199–204.

Valencia, S. (1990). A portfolio approach to classroom reading assessment: The whys, whats, and hows. *The Reading Teacher, 43*, 338–340.

Vogel, S., Hruby, P. J., & Adelman, P. B. (1993). Educational and psychological factors in successful and unsuccessful college students with learning disabilities. *Learning Disability Research & Practice, 8(1)*, 35–43.

Wesson, C. L., & King. R. P. (1992). The role of curriculum-based measurement in portfolio assessment. *Diagnostique, 18(1)*, 27–37.

Whinnery, K. W., & Fuchs, L. S. (1993). Effects of goal and test-taking strategies on the computation performance of students with learning disabilities. *Learning Disability Research & Practice, 8(4)*, 204–214.

Whinnery, K. W., & Stecker, P. M. (1992). Individual progress monitoring to enhance instructional programs in mathematics. *Preventing School Failure, 36(2)*, 26–29.

Wolf, K. (1991). The school teacher's portfolio: Issues in design, implementation, and evaluation. *Phi Delta Kappan, 73(2)*, 130–136.

CHAPTER 6

Armour-Thomas, E., White, M. A., & Boehm, A. (1987, April). *The motivational effects of types of computer feedback on children's learning and retention of relational concepts*. Paper presented at the annual meeting of the American Educational Research Association, Washington, DC. (ERIC Document Reproduction Service No. ED 287 446)

Baker, S. K., Kameenui, E. J., & Simmons, D. C. (in press). The characteristics of students with diverse learning and curricular needs. In E. J. Kameenui & D. W. Carnine (Eds.), *Effective teaching strategies that accommodate diverse learners*. Columbus, OH: Merrill.

Bana, J. P., & Nelson, D. (1977). Some effects of distracters in nonverbal mathematical problems. *Alberta Journal of Educational Research, 23*, 268–279.

Barbetta, P. M., Heward, W. L., Bradley, D. M., & Miller, A. D. (1994). Effects of immediate and delayed error correction on the acquisition and maintenance of sight words by students with developmental disabilities. *Journal of Applied Behavior Analysis, 27*, 177–178.

Baroody, A. J. (1989). Manipulatives don't come with guarantees. *Arithmetic Teacher, 37(2)*, 4–5.

Barron, B., Bransford, J., Kulewicz, S., & Hasselbring, T. (1989, April). *Uses of macrocontexts to facilitate mathematical thinking*. Paper presented at the annual meeting of the American Educational Research Association, San Francisco.

Bereiter, C., & Scardamalia, M. (1987). The psychology of written composition. Hillsdale, NJ: Erlbaum.

Carnine, D. (1980). Preteaching versus concurrent teaching of the component skills of a multiplication algorithm. Journal for Research in Mathematics Education, 11, 375–378.

Carnine, D. (1995). Rational schools: The role of science in helping education become a profession. Behavior and Social Issues, 5(2), 5–19.

Carnine, D., Engelmann, S., Hofmeister, A., & Kelly, B. (1987). Videodisc instruction in fractions. Focus on Learning Problems in Mathematics, 9(1), 31–52.

Carnine, D., Jones, E., & Dixon, R. (1994). Mathematics: Educational tools for diverse learners. School Psychology Review, 23, 406–427.

Carnine, D., & Kameenui, E. (1992). Higher order thinking: Designing curriculum for mainstreamed students. Austin, TX: PRO-ED.

Carnine, D., & Shinn, M. (Eds.). (1994). Introduction to the mini-series: Diverse learners and prevailing, emerging, and research-based educational approaches and their tools. School Psychology Review, 23, 341–471.

Carnine, D., & Stein, M. (1981). Organizational strategies and practice procedures for teaching basic facts. Journal of Research in Mathematics Education, 12, 65–69.

Cawley, J. F., & Miller, J. H. (1989). Cross-sectional comparisons of the mathematical performance of children with learning disabilities: Are we on the right track toward comprehensive programming? Journal of Learning Disabilities, 22, 250–254, 259.

Charles, R. I. (1980). Exemplification and characterization moves in the classroom teaching of geometry concepts. Journal for Research in Mathematics Education, 11, 10–21.

Collins, M., & Carnine, D. (1988). Evaluating the field test revision process by comparing two versions of a reasoning skills CAI program. Journal of Learning Disabilities, 21, 375–379.

Darch, C., Carnine, D., & Gersten, R. (1984). Explicit instruction in mathematics problem solving. Journal of Educational Research, 77, 350–359.

Egan, D. E., & Greeno, J. R. (1973). Acquiring cognitive structure by discovery and rule learning. Journal of Educational Psychology, 64, 85–97.

Engelmann, S., & Carnine, D. (1991). Theory of instruction: Principles and applications. Eugene, OR: Association for Direct Instruction.

Engelmann, S., Carnine, D., Kelly, B., & Engelmann, O. (1991–1995). Connecting math concepts. Chicago: Science Research Associates.

Evans, D. G. (1990). Comparison of three instructional strategies for teaching borrowing in subtraction. Unpublished doctoral dissertation, University of Oregon, Eugene.

Fielding, G., Kameenui, E., & Gersten, R. (1983). A comparison of inquiry and a direct instruction approach to teaching legal concepts and applications to secondary school students. Journal of Educational Research, 76, 287–293.

Fuchs, L. S., Fuchs, D., & Hamlett, C. L. (1989). Computers and curriculum-based measurement: Effects of teacher feedback systems. School Psychology Review, 18, 112–125.

Gelman, R. (1986). *Toward an understanding-based theory of mathematics learning and instruction, or in praise of Lampert on teaching multiplication. Cognition and Instruction, 3,* 349–355.

Gersten, R., & Carnine, D. (1982). *Effective mathematics instruction for low-income students: Results of longitudinal field research in 12 school districts. Journal for Research in Mathematics Education, 13,* 145–152.

Gleason, M., Carnine, D., & Boriero, D. (1990). *Improving CAI effectiveness with attention to instructional design in teaching story problems to mildly handicapped students. Journal of Special Education Technology, 10(3),* 129–136.

Graham, S., & Harris, K. (1988). *Instructional recommendations for teaching writing to exceptional students. Exceptional Children, 54,* 506–512.

Guskey, T. R., & Gates, S. L. (1986). *Synthesis of research on the effects of mastery learning in elementary and secondary classrooms. Educational Leadership, 33(8),* 73–80.

Heibert, J. (1984). *Children's mathematics learning: The struggle to link form and understanding. The Elementary School Journal, 84,* 498–508.

Isaacson, S. (1987). *Effective instruction in written language. Focus on Exceptional Children, 19(6),* 1–12.

Kameenui, E., & Carnine, D. (1986). *Preteaching versus concurrent teaching of the component skills of a subtraction algorithm to skill-deficient second graders: A component analysis of direct instruction. The Exceptional Child, 33(2),* 103–115.

Kameenui, E., Carnine, D., Darch, D., & Stein, M. (1986). *Two approaches to the development phase of mathematics instruction. The Elementary School Journal, 86,* 633–650.

Kelly, B., Carnine, D., Gersten, R., & Grossen, B. (1986). *The effectiveness of videodisc instruction in teaching fractions to learning-disabled and remedial high school students. Journal of Special Education Technology, 8(2),* 5–17.

Kelly, B., Gersten, R., & Carnine D. (1990). *Student error patterns as a function of curriculum design: Teaching fractions to remedial high school students and high school students with learning disabilities. Journal of Learning Disabilities, 23,* 23–29.

Leinhardt, G. (1987). *Development of an expert explanation: An analysis of a sequence of subtraction lessons. Cognition and Instruction, 4,* 225–282.

Lhyle, K. G., & Kulhavy, R. W. (1987). *Feedback processing and error correction. Journal of Educational Psychology, 79,* 320–322.

McDaniel, M. A., & Schlager, M. S. (1990). *Discovery learning and transfer of problem solving. Cognition and Instruction, 7,* 129–159.

McDonnell, J., & Ferguson, G. (1989). *A comparison of time delay and decreasing prompt hierarchy strategies in teaching banking skills to students with moderate handicaps. Journal of Applied Behavior Analysis, 22(1),* 85–91.

Miller, U. C., & Test, D. W. (1989). *A comparison of constant time delay and most-to-least prompting in teaching laundry skills to students with moderate retardation. Education and Training in Mental Retardation, 24,* 363–370.

Montague, M., & Bos, C. S. (1986). *Verbal mathematical problem solving and learning disabilities: A review. Focus on Learning Problems in Mathematics, 8(2),* 7–21.

Moore, L., & Carnine, D. (1989). Evaluating curriculum design in the context of active teaching. Remedial and Special Education, 10(4), 28–37.

Moser, J. M. (1980, April). A longitudinal study of the effect of number six and presence of manipulative materials on children's processes in solving addition and subtraction problems. Washington, DC: ERIC. (ERIC Document Reproduction Service No. ED 184 880)

Mouly, G. J. (1978). Educational research: The art and science of investigation. Boston: Allyn & Bacon.

National Council of Teachers of Mathematics. (1989). Curriculum and evaluation standards for school mathematics. Reston, VA: Author.

Nickerson, R. S. (1985). Understanding. American Journal of Education, 93, 201–239.

Paine, S., Carnine, D., White, W. A. T., & Walters, G. (1982). Effects of fading teacher presentation structure (covertization) on acquisition and maintenance of arithmetic problem-solving skills. Education and Treatment of Children, 5(2), 93–107.

Paris, S. G., Lipson, M. Y., & Wixson, K. K. (1983). Becoming a strategic reader. Contemporary Educational Psychology, 8, 293–316.

Pellegrino, J. W., & Goldman, S. R. (1987). Information processing and elementary mathematics. Journal of Learning Disabilities, 20, 23–32.

Prawat, R. S. (1989). Promoting access to knowledge, strategy, and disposition in students: A research synthesis. Review of Educational Research, 59(1), 1–41.

Prueher, J. (1987). Improving achievement and attitudes in elementary algebra through written error-correcting feedback and free comments on tests. Unpublished master's thesis, National College of Education, Evanston, IL. (ERIC Document Reproduction Service No. ED 282 748)

Resnick, L. (1988). Treating mathematics as an ill-structured discipline. In R. I. Charles & E. A. Silver (Eds.), The teaching and assessing of mathematical problem solving (pp. 32–62). Hillsdale, NJ: Erlbaum.

Resnick, L. B., Cauzinille-Marmeche, E., & Mathieu, J. (1987). Understanding algebra. In J. A. Sloboda & D. Rogers (Eds.), Cognitive processes in mathematics (pp. 169–203). Oxford, England: Clarendon Press.

Resnick, L. B., & Omanson, S. F. (1987). Learning to understand arithmetic. In R. Glaser (Ed.), Advances in instructional psychology (pp. 41–95). Hillsdale, NJ: Erlbaum.

Richman, C. L., & Brown, K. P. (1986). Minimal competency testing: An analysis of three remediation methods. The Journal of Special Education, 20(1), 103–110.

Rosenshine, B., & Stevens, R. (1984). Classroom instruction in reading. In P. D. Pearson (Ed.), Handbook of reading research (pp. 754–798). New York: Longman.

Scardamalia, M., & Bereiter, C. (1986). Research on written composition. In M. Wittrock (Ed.), Handbook on research in teaching (pp. 778–803). New York: Macmillan.

Schimmel, B. J. (1983, April). A meta-analysis of feedback to learners in computerized and programmed instruction. Paper presented at the annual meeting of the American Educational Research Association, Montreal, Canada. (ERIC Document Reproduction Service No. ED 233 708)

Siegel, P. S., & Crawford, K. A. (1983). *Two-year follow-up study of discrimination learning by mentally retarded children. American Journal of Mental Deficiency, 88(1),* 76–78.

Stein, M., Silbert, J., & Carnine, D. (in press). *Direct instruction mathematics.* Columbus, OH: Merrill.

Stigler, J. W., & Baranes, R. (1988). *Culture and mathematics learning. Review of Research in Education, 15,* 253–306.

Tharp, R. G., & Gallimore, R. (1988). *Rousing minds to life: Teaching, learning, and schooling in social context.* New York: Cambridge University Press.

Thomas, C., Englert, C. S., & Gregg, S. (1987). *An analysis of errors and strategies in the expository writing of learning disabled students. Remedial and Special Education, 8(1),* 21–30.

Trafton, P. R. (1984). *Toward more effective, efficient instruction in mathematics. The Elementary School Journal, 84,* 514–530.

Woodward, J. (1994). *Effects of curriculum discourse style on eighth graders' recall and problem solving in earth science. The Elementary School Journal, 94,* 299–314.

Woodward, J., Carnine, D., & Gersten, R. (1988). *Teaching problem solving through computer simulation. American Educational Research Journal, 25(1),* 72–86.

CHAPTER 7

Baroody, A. J., & Hume, J. (1991). *Meaningful mathematics instruction: The case of fractions. Remedial and Special Education, 12(3)* 54–68.

Behrend, J. L. (1994). *Mathematical problem-solving processes of primary-grade students identified as learning disabled.* Unpublished doctoral dissertation, University of Wisconsin, Madison.

Bley, N. S., & Thornton, C. A. (1994). *Teaching mathematics to students with learning disabilities.* Austin, TX: PRO-ED.

Borasi, R. (1996). *The realities, challenges and promise of teaching mathematics to all students.* In D. Schifter (Ed.), *Voicing the new pedagogy: Classroom narratives and the construction of meaning for the rhetoric of mathematics education reform* (Vol. 1). New York: Teachers College Press.

Borasi, R., Kort, E., Leonard, P., & Stone, G. (1993). *Report of an implementation of a remodeling unit in a class of learning disabled students.* In R. Borasi & C. F. Smith (Eds.), *Remodeling: Experiencing mathematics within a real-life context—Detailed reports of three classroom implementations* (pp. 138–166) (Interim Report, Supporting Middle School Learning Disabled Students in the Mainstream Mathematics Classroom, Project No. TPE-9153812). Arlington, VA: National Science Foundation.

Borasi, R., Packman, D., & Woodward, A. (1991). *Supporting middle school learning disabled students in the mainstream mathematics classroom* (Project No. TPE-9153812). Arlington, VA: National Science Foundation.

Bulgren, J., & Montague, M. (1989, June). *Report from working group four*. Paper presented at the Information Center for Special Education Media and Materials Instructional Methods Forum, Washington, DC.

Carpenter, T. P., Fennema, E., Peterson, P. L., Chiang, C.P., & Loef, M. (1989). Using knowledge of children's mathematics thinking in classroom teaching: An experimental study. *American Education Research Journal, 26*, 499–532.

Carpenter, T. P., & Moser, J. M. (1984). The acquisition of addition and subtraction concepts in grades one through three. *Journal for Research in Mathematics Education, 15*, 179–202.

Cawley, J., Fitzmaurice-Hayes, A., & Shaw, R. (1988). *Mathematics for the mildly handicapped—A guide to curriculum and instruction*. Boston: Allyn & Bacon.

Cawley, J., & Miller, J. (1989). Cross-sectional comparisons of the mathematical performance of children with learning disabilities: Are we on the right track toward comprehensive programming? *Journal of Learning Disabilities, 22*, 250–254.

Cobb, P., & Bauersfeld, H. (1995). *The emergence of mathematical meaning: Interaction in classroom cultures*. Hillsdale, NJ: Erlbaum.

Cobb, P., Wood, T., Yackel, E., Nicholls, J., Wheatley, G., Trigatti, B., & Perlwitz, M. (1991). Assessment of a problem-centered second-grade mathematics project. *Journal for Research in Mathematics Education, 22(1)*, 3–29.

Englert, C. S., Tarrant, K. L., & Mariage, T. V. (1992). Defining and redefining instructional practice in special education: Perspectives on good teaching. *Teacher Education and Special Education, 15(2)*, 62–86.

Fennema, E., & Carpenter, T. P. (1985). *Cognitively guided instruction: A program implementation guide*. Madison: Wisconsin Center for Education Research.

Grouws, D. A. (Ed.). (1992). *Handbook of research on mathematics teaching and learning*. New York: Macmillan.

Heshusius, L. (1991). Curriculum-based assessment and direct instruction: Critical reflections on fundamental assumptions. *Exceptional Children, 57*, 315–328.

Jones, G. A., Thornton, C. A., Putt, I. J., Hill, K. M., Mogill, A. T., Rich, B. S., & van Zoest, L. R. (1996). Multidigit number sense: A framework for instruction and assessment. *Journal for Research in Mathematics Education, 27*, 310–336.

Langrall, C. W., Thornton, C. A., Jones, G. A., & Malone, J. A. (1996). Enhanced pedagogical knowledge and reflective analysis in elementary mathematics teacher education. *Journal of Teacher Education, 47*, 271–282.

Lo, J.-J., Wheatley, G. H., & Smith, A. D. (1991, April). *Learning to talk mathematics*. Paper presented at the annual meeting of the American Education Research Association, Chicago.

Marshall, H. H. (1988). Work or learning; Implications of classroom metaphors. *Educational Researcher, 17*, 9–16.

Mastropieri, M. A., Scruggs, T. E., & Shiah, S. (1991). Mathematics instruction for learning disabled students: A review of research. *Learning Disabilities Research & Practice, 6*, 89–98.

McTighe, J., & Lyman, F. T., Jr. (1988). Cueing thinking in the classroom: The promise of theory-embedded tools. Educational Leadership, 45(7), 18–24.

Montague, M., & Bos, C. (1986). Verbal mathematical problem solving and learning disabilities: A review. Focus on Learning Problems in Mathematics, 8, 7–21.

National Council of Teachers of Mathematics. (1989). Curriculum and evaluation standards for school mathematics. Reston, VA: Author.

National Council of Teachers of Mathematics. (1991). Professional standards for teaching mathematics. Reston, VA: Author.

National Council of Teachers of Mathematics. (1995). Assessment standards for school mathematics. Reston, VA: Author.

National Research Council. (1990). Reshaping school mathematics. Washington, DC: National Academy Press.

Noddings, N. (1990). Constructivism in mathematics education. In R. B. Davis, C. A. Maher, & N. Noddings (Eds.), Constructivist views on the teaching and learning of mathematics [Monograph]. Journal for Research in Mathematics Education, 4, 7–18.

Pimm, D. (1987). Speaking mathematically: Communication in mathematics classrooms. New York. Routledge & Kegan Paul.

Resnick, L. B. (1987). Education and learning to think. Washington, DC: National Academy Press.

Resnick, L. B. (1989). Developing mathematical knowledge. American Psychologist, 44, 162–169.

Scheid, K. (1990). Cognitive-based methods for teaching mathematics to students with learning problems. Columbus, OH: Information Center for Special Education Media and Materials.

Silver, E. G. (Ed.). (1985). Teaching and learning mathematical problem solving: Multiple research perspectives. Hillsdale, NJ: Erlbaum.

Speer, W. R., & Brahier, D. J. (1994). Rethinking the teaching and learning of mathematics. In C. A. Thornton & N. S. Bley (Eds.), Windows of opportunity: Mathematics for students with special needs (pp. 41–59). Reston, VA: National Council of Teachers of Mathematics.

Stenmark, J. (1991). Mathematics assessment: Myths, models and practical suggestions. Reston, VA: National Council of Teachers of Mathematics.

Stone, G. (1993). Report of an implementation of an area unit in a class of learning disabled students. In R. Borasi (Ed.), Developing area formulas: An opportunity for inquiry within the traditional math curriculum—Detailed reports of three classroom implementations (pp. 42–87) (Interim Report, Supporting Middle School Learning Disabled Students in the Mainstream Mathematics Classroom, Project No. TPE-9153812). Arlington, VA: National Science Foundation.

Thornton, C. A., & Bley, N. S. (1994). Windows of opportunity: Mathematics for students with special needs. Reston, VA: National Council of Teachers of Mathematics.

Trafton, P. R., & Claus, A. S. (1994). A changing curriculum for a changing age. In C. A. Thornton & N. S. Bley (Eds.), Windows of opportunity: Mathematics for *students with special needs (pp. 19–39). Reston, VA: National Council of Teachers of Mathematics.*

Wansart, W. (1990). *Learning to solve a problem: A microanalysis of the solution strategies of children with learning disabilities. Journal of Learning Disabilities, 23,* 164–170.

CHAPTER 8

Algozzine, B., O'Shea, D. J., Crews, W. B., & Stoddard, K. (1987). *Analysis of mathematics competence of learning disabled adolescents. The Journal of Special Education, 21,* 97–107.

Archer, A., & Isaacson, S. (1989). *Design and delivery of academic instruction. Reston, VA: Council for Exceptional Children.*

Becker, W. C., & Carnine, D. W. (1981). *Direct instruction: A behavior theory model for comprehensive educational intervention with the disadvantaged. In S. W. Bijou & R. Ruiz (Eds.), Behavior modification: Contributions to education (pp.* 145–210). *Hillsdale, NJ: Erlbaum.*

Bereiter, C. (1994). *Constructivism, socioculturalism, and Popper's World 3. Educational Researcher, 23(7),* 21–23.

Bottge, B. A., & Hasselbring, T. S. (1993). *A comparison of two approaches for teaching complex, authentic mathematics problems to adolescents in remedial math classes. Exceptional Children, 59,* 556–566.

Bransford, J. D. , Sherwood, R. D., Hasselbring, T. S., Kinzer, C. K., & Williams, S. M. (1990). *Anchored instruction: Why we need it and how technology can help. In D. Nix & R. Spiro (Eds.), Cognition, education, multimedia (pp. 115–141). Hillsdale, NJ: Erlbaum.*

Brophy, J. E., & Good, T. L. (1974). *Teacher–student relationships: Causes and consequences. New York: Holt, Rinehart &Winston.*

Carnine, D. (1980). *Preteaching vs. concurrent teaching of the component skills of a multiplication algorithm. Journal for Research in Mathematics Education, 11,* 375–379.

Carnine, D. (1989). *Designing practice activities. Journal of Learning Disabilities, 22,* 603–607.

Carnine, D. (1992). *Expanding the notion of teachers' rights: Access to tools that work. Journal of Applied Behavior Analysis, 25,* 13–19.

Carpenter, R. L. (1985). *Mathematics instruction in resource rooms. Learning Disability Quarterly, 8,* 95–100.

Carpenter, T. P., Matthews, W., Linquist, M. M., & Silver, E. A. (1984). *Achievement in mathematics: Results from the national assessment. Elementary School Journal, 84,* 485–495.

Case, L. P., Harris, K. R., & Graham, S. (1992) *Improving the mathematical problem-solving skills of students with learning disabilities: Self-regulated strategy development. The Journal of Special Education, 26,* 1–19.

Cawley, J. F., Fitzmaurice, A. M., Shaw, R., Kahn, H., & Bates, H., III. (1978). *Mathematics and learning disabled youth: The upper grade levels.* Learning Disability Quarterly, 1(4), 37–52.

Cawley, J. F., Fitzmaurice, A. M., Shaw, R., Kahn, H., & Bates, H., III. (1979). *LD youth and mathematics: A review of characteristics.* Learning Disability Quarterly, 2(1), 29–44.

Cawley, J. F., & Miller, J. H. (1989). *Cross-sectional comparisons of the mathematical performance of children with learning disabilities: Are we on the right track toward comprehensive programming?* Journal of Learning Disabilities, 22, 250–259.

Chapman, J. (1988). *Learning disabled children's self-concepts.* Review of Educational Research, 58, 347–371.

Chow, S. (1981). *A study of academic learning time of mainstreamed LD students.* San Francisco: Far West Laboratory for Educational Research.

Christenson, S. L., Ysseldyke, J. E., & Thurlow, M. (1989). *Critical instructional factors for students with mild handicaps: An integrative review.* Remedial and Special Education, 10(5), 21–31.

Cobb, P. (1988). *The tension between theories of learning and instruction in mathematics education.* Educational Psychologist, 23, 87–103.

Cobb, P. (1994a). *Constructivism in mathematics and science education.* Educational Researcher, 23(7), 4.

Cobb, P. (1994b). *Where is the mind? Constructivist and sociocultural perspectives on mathematical development.* Educational Researcher, 23(7), 13–20.

Cognition and Technology Group at Vanderbilt University. (1991). *Technology and design of generative learning environments.* Educational Technology, 3(5), 34–40.

Cybriwsky, C. A., & Schuster, J. W. (1990). *Using constant delay procedures to teach multiplication facts.* Remedial and Special Education, 11(1), 54–59.

Darch, C., Carnine, D., & Gersten, R. (1984). *Explicit instruction in mathematics problem solving.* Journal of Educational Research, 77, 350–359.

Deshler, D. D., & Schumaker, J. B. (1986). *Learning strategies: An instructional alternative for low achieving adolescents.* Exceptional Children, 52, 583–590.

Driver, R., Asoko, H., Leach, J., Mortimer, E., & Scott, P. (1994). *Constructing scientific knowledge in the classroom.* Educational Researcher, 23(7), 5–12.

Ellis, E. S., Lenz, B. K., & Sabornie, E. J. (1987a). *Generalization and adaptation of learning strategies to natural environments: Part 1. Critical agents.* Remedial and Special Education, 8(1), 6–20.

Ellis, E. S., Lenz, B. K., & Sabornie, E. J. (1987b). *Generalization and adaptation of learning strategies to natural environments: Part 2. Research into practice.* Remedial and Special Education, 8(2), 6–23.

Engelmann, S. (1993). *Priorities and efficiency.* LD Forum, 18(2), 5–8.

Engelmann, S., & Carnine, D. (1982). *Theory of instruction: Principles and applications.* New York: Irvington.

Fuchs, L. S., & Deno, S. L. (1991). *Paradigmatic distinctions between instructionally relevant measurement models.* Exceptional Children, 57, 488–499.

Fuchs, L. S., Fuchs, D., & Deno, S. L. (1985). *Importance of goal ambitiousness and goal mastery to student achievement.* Exceptional Children, 52, 63–71.

Fuchs, L. S., Fuchs, D., Hamlett, C. L., & Stecker, P. M. (1990).The role of skills analysis in curriculum-based measurement in math. School Psychology Review, 19, 6–22.

Gadanidis, G. (1994). Deconstructing constructivism. The Mathematics Teacher, 87(2), 91–97.

Gersten, R. (1985). Direct instruction with special education students: A review of evaluation research. The Journal of Special Education, 19, 41–58.

Harris, K. R., & Graham, S. (1994). Constructivism: Principles, paradigms, and integration. The Journal of Special Education, 28, 233–247.

Hasselbring, T. S., Goin, L. T., & Bransford, J. D. (1988). Developing math automaticity in learning handicapped children: The role of computerized drill and practice. Focus on Exceptional Children, 20(6), 1–7.

Hasselbring, T. S., Sherwood, R. D., Bransford, J. D., Mertz, J., Estes, B., Marsh, J., & Van Haneghan, J. (1991). An evaluation of specific videodisc courseware on student learning in a rural school environment (Final Report). Nashville: TN: Peabody College of Vanderbilt University, The Learning Technology Center. (ERIC Document Reproduction Service No. ED 338 225)

Hutchinson, N. L. (1993). The effect of cognitive instruction on algebra problem solving of adolescents with learning disabilities. Learning Disability Quarterly, 16, 34–63.

Jitendra, A. K., Kameenui, E., & Carnine, D. (1994). An exploratory evaluation of dynamic assessment of the role of basals on comprehension of mathematical operations. Education and Treatment of Children, 17, 139–162

Kameenui, E. J., & Simmons, D. C. (1990). Designing instructional strategies: The prevention of academic problems. Columbus, OH: Merrill.

Kelly, B., Gersten, R., & Carnine, D. (1990). Student error patterns as a function of curriculum design: Teaching fractions to remedial high school students and high school students with learning disabilities. Journal of Learning Disabilities, 23, 23–29

Koscinski, S. T., & Gast, D. L. (1993). Use of constant time delay in teaching multiplication facts to students with learning disabilities. Journal of Learning Disabilities, 26, 533–544, 567.

Koscinski, S. T., & Hoy, C. (1993). Teaching multiplication facts to students with learning disabilities: The promise of constant time delay procedures. Learning Disabilities Research and Practice, 8, 260–263.

Lenz, B. K., & Deshler, D. D. (1990). Principles of strategies instruction as the basis of effective preservice teacher education. Teacher Education and Special Education, 13, 82–95.

Madden, N. A., & Slavin, R. E. (1983). Effects of cooperative learning on the social acceptance of mainstreamed academically handicapped students. The Journal of Special Education, 17, 171–182.

Maheady, L., Sacca, M. K., & Harper, G. F. (1987). Classwide student tutoring teams: The effects of peer-mediated instruction on the academic performance of secondary mainstreamed students. The Journal of Special Education, 21, 107–121.

McLeod, T. M., & Armstrong, S. W. (1982). *Learning disabilities in mathematics— Skill deficits and remedial approaches at the intermediate and secondary level.* Learning Disability Quarterly, 5, 305–311.

McLoone, B., Scruggs, T., Mastropieri, M., & Zucker, S. (1986). *Memory strategy instruction and training with LD adolescents.* Learning Disabilities Research, 2(1), 45–53.

Mercer, C. D., Jordan, L., & Miller, S. P. (1994). *Implications of constructivism for teaching math to students with moderate to mild disabilities.* The Journal of Special Education, 28, 290–306.

Mercer, C. D., & Miller, S. P. (1992). *Teaching students with learning problems in math to acquire, understand, and apply basic math facts.* Remedial and Special Education, 13(3), 19–35, 61.

Montague, M. (1992). *The effects of cognitive and metacognitive strategy instruction on the mathematical problem solving of middle school students with learning disabilities.* Journal of Learning Disabilities, 25, 230–248.

Montague, M., & Bos, C. S. (1986). *The effect of cognitive strategy training on verbal math problem solving performance of learning disabled adolescents.* Journal of Learning Disabilities, 19, 26–33.

Montague, M., Bos, C., & Doucette, M. (1991). *Affective, cognitive, and metacognitive attributes of eighth-grade mathematical problem solvers.* Learning Disabilities: Research & Practice, 6, 145–151.

National Council of Teachers of Mathematics. (1989). *Curriculum and evaluation standards for school mathematics.* Reston, VA: Author.

Nesher, P. (1976). *Three determinants of difficulty in verbal arithmetic problems.* Educational Studies in Mathematics, 7, 369–388.

Pajares, F., & Miller, M. D. (1994). *Role of self-efficacy and self-concept beliefs in mathematical problem solving: A path analysis.* Journal of Educational Psychology, 86, 193–203.

Perkins, V., & Cullinan, D. (1985). *Effects of direct instruction for fraction skills.* Education and Treatment of Children, 8, 41–50.

Poplin, M. S. (1988a). *Holistic/constructivist principles of the teaching/learning process: Implications for the field of learning disabilities.* Journal of Learning Disabilities, 21, 401–416.

Poplin, M. S. (1988b). *The reductionistic fallacy in learning disabilities: Replicating the past by reducing the present.* Journal of Learning Disabilities, 21, 389–400.

Pressley, M., Harris, K. R., & Marks, M. R. (1992). *But good strategy instructors are constructivists!* Educational Psychology Review, 4, 3–31.

Pressley, M., Symons, S., Snyder, B. L., & Cariglia-Bull, T. (1989). *Strategy instruction research comes of age.* Learning Disability Quarterly, 12, 16–30.

Reid, D. K., Kurkjian, C., & Carruthers, S. S. (1994). *Special education teachers interpret constructivist teaching.* Remedial and Special Education, 15, 267–280.

Reith, H., & Evertson, C. (1988). *Variables related to the effective instruction of difficult-to-teach children.* Focus on Exceptional Children, 20(5), 1–8.

Reschly, D. J. (1992). Special education decision making and functional/behavioral assessment. In W. Stainback & S. Stainback (Eds.), Controversial issues confronting special education: Divergent perspectives (pp. 127–138). Boston: Allyn & Bacon.

Rivera, D., & Smith, D. D. (1988). Using a demonstration strategy to teach midschool students with learning disabilities how to compute long division. Journal of Learning Disabilities, 21, 77–81.

Rosenshine, B. (1976). Classroom instruction. In N. L. Gage (Ed.), The psychology of teaching methods. Chicago: University of Chicago Press.

Rosenshine, B., & Stevens, R. (1986). Teaching functions. In M. C. Whittrock (Ed.), Handbook of research on teaching (3rd ed., pp. 376–391).

Silbert, J., Carnine, D., & Stein, M. (1990). Direct instruction mathematics (2nd ed.). Columbus, OH: Merrill.

Slavin, R. E. (1983). When does cooperative learning increase student achievement? Psychological Bulletin, 94, 429–445.

Slavin, R. E. (1984). Team assisted individualization: Cooperative learning and individualized instruction in the mainstreamed classroom. Remedial and Special Education, 5(6), 33–42.

Slavin, R. E., & Karweit, N. L. (1981). Cognitive and affective outcomes of an intensive student team learning experience. Journal of Experimental Education, 50, 29–35.

Slavin, R. E., & Karweit, N. L. (1984). Mastery learning and student teams: A factorial experiment in urban general mathematics classes. American Journal of Educational Research, 21, 725–736.

Slavin, R. E., Leavey, M., & Madden, N. A. (1984). Combining cooperative learning and individualized instruction: Effects on student mathematics achievement, attitudes, and behaviors. Elementary School Journal, 84, 409–422.

Slavin, R. E., Madden, N. A., & Leavey, M. (1984). Effects of team assisted individualization on the mathematics achievement of academically handicapped and nonhandicapped students. Journal of Educational Psychology, 76, 813–819.

Stevens, K. B., & Schuster, J. W., (1988). Time delay: Systematic instruction for academic tasks. Remedial and Special Education, 9(5), 16–21.

Swing, S. R., Stoiber, K. C., & Peterson, P. L. (1988). Thinking skills vs. learning time: Effects of alternative classroom-based interventions on students mathematics problem solving. Cognition and Instruction, 5, 123–191.

Tennyson, R. D., & Park, O. (1980). The teaching of concepts: A review of instructional design research literature. Review of Educational Research, 50, 55–70.

Woodward, J. (1991). Procedural knowledge in mathematics: The role of the curriculum. Journal of Learning Disabilities, 24, 242–251.

Wright, J. P. (1968). A study of children's performance on verbally stated problems containing word clues and omitting them. Dissertation Abstracts International, 29, 1770B.

Zigmond, N. (1990). Rethinking secondary school programs for students with learning disabilities. Focus on Exceptional Children, 23(1), 2–22.

CHAPTER 9

Alexander, P,. & Judy, J. (1988). The interaction of domain-specific and strategic knowledge in academic performance. Review of Educational Research, 58, 375–404.

Anderson, J. (1990). The adaptive character of thought. Hillsdale, NJ: Erlbaum.

Bennett, K. (1982). The effects of syntax and verbal mediation on learning disabled students' verbal mathematical problem solving (Doctoral dissertation, Northern Arizona University, 1981). Dissertation Abstracts International, 42, 422–429.

Biggs, J. B., & Collis, K. F. (1982). Evaluating the quality of learning: The SOLO taxonomy. New York: Academic Press.

Case, L., Harris, K., & Graham, S. (1992). Improving the mathematical problem solving skills of students with learning disabilities: Self-instructional strategy development. The Journal of Special Education, 21, 1–19.

Case, R. (1985). Intellectual development: Birth to adulthood. Orlando, FL: Academic Press.

Case, R., Hayward, S., Lewis, M., & Hurst, P. (1988). Toward a neo-Piagetian theory of cognitive and emotional development. Developmental Review, 8, 1–51.

Ellis, E. S. (1993). Integrative strategy instruction: A potential model for teaching content area subjects to adolescents with learning disabilities. Journal of Learning Disabilities, 26, 358–383.

Englert, C. S. (1993). Writing instruction from a sociocultural perspective: The holistic, dialogic, and social enterprise of writing. Journal of Learning Disabilities, 25, 153–172.

Flavell, J. (1985). Cognitive development (2nd ed.). Englewood Cliffs, NJ: Prentice-Hall.

Geary, D. C. (1993). Mathematical disabilities: Cognitive, neuropsychological. and genetic components. Psychological Bulletin, 114, 345–362.

Ginsburg, H. P., Jacobs, S., & Lopez, L. (1993). Assessing mathematical thinking and learning potential. In R. B. Davis & C. A. Maher (Eds.), Schools, mathematics, and the world of reality (pp. 237–262). Boston: Allyn & Bacon.

Ginsburg, H. P., & Russell, R. L. (1981). Social class and racial influences on early mathematical thinking. Monographs of the Society for Research in Child Development, 46 (6, Serial No. 193).

Graham, S., & Harris, K. R. (1994). The role and development of self-regulation in the writing process. In D. Schunk & B. Zimmerman (Eds.), Self-regulation of learning and performance: Issues and educational applications (pp. 203–228) Hillsdale, NJ: Erlbaum.

Groteluschen, A. K., Borkowski, J. G., & Hales. C. (1990). Strategy instruction is often insufficient: Addressing the interdependency of executive and attributional processes. In T. Scruggs & B. Wong (Eds.), Intervention research in learning disabilities (pp. 81–101). New York: Springer-Verlag.

Hiebert, J. & Carpenter, T. (1992). Learning and teaching with understanding. In D. Grouws (Ed.), Handbook of research on mathematics teaching and learning (pp. 65–97). New York: Macmillan.

Huntington, D. (1994). *Instruction in concrete, semi-concrete, and abstract representation as an aid to the solution of relational problems by adolescents with learning disabilities.* Unpublished doctoral dissertation, University of Georgia, Athens.

Hutchinson, N. (1993). Effects of cognitive strategy instruction on algebra problem solving of adolescents with learning disabilities. *Learning Disability Quarterly, 16,* 34–63.

Jastak, S., & Wilkinson, G. (1984). *Wide range achievement test-Revised.* Wilmington, DE: Jastak & Assoc.

Kolligian, J., & Sternberg, R. (1987). Intelligence, information processing, and learning disabilities: A triarchic synthesis. *Journal of Learning Disabilities, 20,* 8–17.

Mandler, G. (1989). Affect and learning: Causes and consequences of emotional interactions. In D. B. McLeod & V. M. Adams (Eds.), *Affect and mathematical problem solving: A new perspective* (pp. 3–19). Hillsdale, NJ: Erlbaum.

Marzola, E. S. (1987, April). *An arithmetic verbal problem-solving model far learning disabled students.* Paper presented at the annual meeting of the American Educational Research Association, Washington, DC.

Montague, M. (1992). The effects of cognitive and metacognitive strategy instruction on mathematical problem solving of middle school students with learning disabilities. *Journal of Learning Disabilities, 25,* 230–248.

Montague, M., & Applegate, B. (1993a). Mathematical problem-solving characteristics of middle school students with learning disabilities. *The Journal of Special Education, 27,* 175–201.

Montague, M., & Applegate. B. (1993b). Middle school students' mathematical problem solving: An analysis of think-aloud protocols. *Learning Disability Quarterly, 16,* 19–32.

Montague, M., Applegate, B., & Marquard, K. (1993). Cognitive strategy instruction and mathematical problem-solving performance of students with learning disabilities. *Learning Disabilities Research & Practice, 8,* 223–232.

Montague, M., & Bos, C. (1986). The effect of cognitive strategy training on verbal math problem solving performance of learning disabled adolescents. *Journal of Learning Disabilities, 19,* 26–33.

Montague, M., & Bos, C. (1990). Cognitive and metacognitive characteristics of eighth-grade students' mathematical problem solving. *Learning and Individual Differences, 2,* 109–127.

Montague, M., Bos, C., & Doucette, M. (1991). Affective, cognitive. and metacognitive attributes of eighth-grade mathematical problem solvers. *Learning Disabilities Research & Practice, 6,* 145–151.

Montague, M., Marquard, K., & LeBlanc, W. (1993, April). *The effects of cognitive strategy instruction on the mathematical problem-solving of students with learning disabilities.* Paper presented at the annual meeting of the American Educational Research Association, Atlanta, GA.

National Council of Teachers of Mathematics. (1989). *Curriculum and evaluation standards for school mathematics.* Reston, VA: Author.

Nuzum, M. (1983). Teaching the arithmetic story problem process. *Journal of Reading, Writing, and Learning Disabilities International, 3,* 53–61.

Piaget, J. (1952). The origins of intelligence in children. New York: International Universities Press.

Pogrow, S. (1992). A validated approach to thinking development for at-risk populations. In C. Collins & J. N. Mangieri (Eds.), Teaching thinking: An agenda for the 21st century (pp. 87–101). Hillsdale, NJ: Erlbaum.

Pressley, M., & Associates. (1990). Cognitive strategy instruction that really improves children's academic performance. Cambridge, MA: Brookline.

Reid, M. K., & Borkowski, J. G. (1987). Causal attributions of hyperactive children: Implications for teaching strategies and self control. Journal of Educational Psychology, 79, 296–307.

Saxe, G. B. (1991). Culture and cognitive development: Studies in mathematical understanding. Hillsdale, NJ: Erlbaum.

Scruggs, T., & Mastropieri, M. (1993). Special education for the twenty-first century: Integrating learning strategies and thinking skills. Journal of Learning Disabilities, 26, 392–398.

Scruggs, T., & Wong, B. (1990). Intervention research in learning disabilities. New York: Springer-Verlag.

Siegler, R. S., & Jenkins, E. (1989). How children discover strategies. Hillsdale, NJ: Erlbaum.

Smith, E., & Alley, G. (198)). The effect of teaching sixth graders with learning difficulties a strategy for solving verbal math problems (Research Report No. 39). Lawrence: The University of Kansas Institute for Research in Learning Disabilities.

Sternberg, R. J. (1985). Beyond IQ: A triarchic theory of human intelligence. Cambridge, England: Cambridge University Press.

Swanson, H. L. (1990). Instruction derived from the strategy deficit model: Overview of principles and procedures. In T. Scruggs & B. Wong (Eds.), Intervention research in learning disabilities (pp. 34–65). New York: Springer-Verlag.

Vauras, M., Lehtinen, E., Olkinuora, E., & Salonen, P. (1993). Devices and desires: Integrative strategy instruction from a motivational perspective. Journal of Learning Disabilities, 26, 384–391.

Vygotsky, L. (1986). Thought and language. Cambridge, MA: MIT Press.

Wong, B. Y. L. (1992). On cognitive process-based instruction: An introduction. Journal of Learning Disabilities, 25, 150–152.

Wong, B. Y. L. (1993). Pursuing an elusive goal: Molding strategic teachers and learners. Journal of Learning Disabilities, 26, 354–357.

Wong, B. Y. L. (1994). Instructional parameters promoting transfer of learned strategies in students with learning disabilities. Learning Disability Quarterly, 17, 110–120.

Woodcock, R., & Johnson, W. (1989). Woodcock-Johnson psycho-educational battery. Boston: Teaching Resources Corp.

Zawaiza, T., & Gerber, M. (1993). Effects of explicit instruction on math word-problem solving by community college students with learning disabilities. Learning Disability Quarterly, 16, 64–79.

Zentall, S. S., & Ferkis, M. A. (1993). Mathematical problem solving for youth with ADHD, with and without learning disabilities. Learning Disability Quarterly, 16, 6–18.

CHAPTER 10

Barbieri, M. M. & Wircenski, J. L. (1990). Developing integrated curricula. Journal for Vocational Special Needs Education, 13, 27–29.

Chambers, D. L., & Kepner, H. S. (1985). Applied mathematics: A three-year program for non–college-bound students. In C. R. Hirsch & M. J. Zweng (Eds.), The secondary school mathematics curriculum (pp. 211–229). Reston, VA: National Council of Teachers of Mathematics.

Clark, G. M., Field, S., Patton, J. P., Brolin, D. E., & Sitlington, P. L. (1994). Life skills instruction: A necessary component for all students with disabilities (a position statement of the Division on Career Development and Transition). Career Development for Exceptional Individuals, 17, 125–133.

Cronin, M. E., & Patton, J. R. (1993). Life skills instruction for all students with special needs: A practical guide for integrating real-life content into the curriculum. Austin, TX: PRO-ED.

Helmke, L., Havekost, D. M., Patton, J. R., & Polloway, E. A. (1994). Life skills programming: Development of a high school science course. Teaching Exceptional Children, 26(2), 49–53.

Hofmeister, A. (1993). Elitism and reform in school mathematics. Remedial and Special Education, 14, 8–13.

Hutchinson, N. (1993). Students with disabilities and mathematics education reform— Let the dialogue begin. Remedial and Special Education, 14, 20–23.

Lindsay, E. R. (1979). This is your life: An applied mathematics curriculum for young adults. In S. Sharron & R. E. Reys (Eds.), Applications in school mathematics (1979 Yearbook). Reston, VA: National Council of Teachers of Mathematics.

Marzano, R. L., & Kendall, J. S. (1995). The McREL database: A tool for constructing local standards. Educational Leadership, 52, 42–47.

Mercer, C. D., Harris, C. A., & Miller, S. P. (1993). Reforming reforms in mathematics. Remedial and Special Education, 14, 14–19.

Murphy, S., & Walsh, J. (1989) Economics and the real life connections. Social Studies and Young Learner, 2, 6–8.

National Council of Teachers of Mathematics. (1980). Agenda for action. Reston, VA: Author.

National Council of Teachers of Mathematics. (1989). Curriculum and evaluation standards for school mathematics. Reston, VA: Author.

Pickard, S. (1990). Integrating math skills into vocational education curricula. Journal for Vocational Special Needs Education, 13, 9–13.

Polloway, E. A., Patton, J. R., Epstein, M. H., & Smith, T. (1989). Comprehensive curriculum for students with mild handicaps. Focus on Exceptional Children, 21(8), 1–12.

Rivera, D. M. (1993). Examining mathematics reform and the implications for students with mathematics disabilities. Remedial and Special Education, 14, 24–27.

Sonnabend, T. (1985). Noncareer mathematics: The mathematics we all need. In C. R. Hirsch & M. J. Zweng (Eds.), The secondary school mathematics curriculum (pp. 107–118). Reston, VA: National Council of Teachers of Mathematics.

Vatter, T. (1994). Civic mathematics: A real-life general mathematics course. The Mathematics Teacher, 87, 396–401.

Wagner, M., Blackorby, J., Cameto, R., Hebbeler, K., & Newman, L. (1993). The transition experiences of young people with disabilities: Findings from the National Longitudinal Transition Study of Special Education Students. Menlo Park, CA: SRI International.

Wagner, M., Newman, L., D'Amico, R., Jay, E. D., Butler-Nalin, P., Marder, C., & Cox, R. (1991). Youth with disabilities: how are they doing? Menlo Park, CA: SRI International.

CHAPTER 11

Algozzine, B., O'Shea, D., Crews, W. B., & Stoddard, K. (1987). Analysis of mathematics competence of learning disabled adolescents. The Journal of Special Education, 21(2), 97–107.

Bruner, J. S. (1966). Toward a theory of instruction. Cambridge, MA: Harvard University Press.

Carnine, D. (1991). Curricular interventions for teaching higher order thinking to all students: Introduction to the special series. Journal of Learning Disabilities, 24, 261–269.

Cawley, J. F., Fitzmaurice, A. M., Goodstein, H., Kahn, H., & Bates, H. (1979). LD youth and mathematics: A review of characteristics. Learning Disability Quarterly, 2(1), 29–44.

Cawley, J. F., & Goodman, J. G. (1968). Interrelationships among mental abilities, reading language arts and arithmetic with the midly handicapped. Arithmetic Teacher, 15, 74–89.

Cawley, J. F., Kahn, H., & Tedesco, A. (1989). Vocational education and students with learning disabilities. Journal of Learning Disabilities, 23, 284–290.

Cawley, J. F., & Miller, J. H. (1989). Cross-sectional comparisons of the mathematical performance of learning disabled children: Are we on the right track toward comprehensive programming? Journal of Learning Disabilities, 22, 250–255.

Cawley, J. F., & Parmar, R. S. (1992). Arithmetic programming for students with disabilities: An alternative. Remedial and Special Education, 13(3), 6–18.

Cawley, J. F., Parmar, R. S., & Smith, M. A. (1995). An analysis of the performance of students with mild disabilities on KEYMATH and KEYMATH-R. Unpublished manuscript, State University of New York at Buffalo.

Chandler, D. G., & Brosnan, P. A. (1994). Mathematics textbook changes from before to after 1989. Focus on Learning Problems in Mathematics, 16, 2–10.

Cobb, P. (1994). Where is the mind? Constructivist and sociocultural perspectives on mathematical development. Educational Researcher, 23(7), 13–20.

Cosden, M. (1988). Microcomputer instruction and perceptions of effectiveness by special education and regular elementary school teachers. The Journal of Special Education, 22, 242–253.

Cox, L. (1975). Systematic errors in the four vertical algorithms in normal and handi-capped populations. Journal of Research in Mathematics Education, 6, 202–220.

Fennema, E., & Franke, M. L. (1992). Teachers' knowledge and its impact. In D. A. Grouws (Ed.), Handbook of research on mathematics teaching and learning (pp. 147–164). New York: Macmillan.

Graves, A., Landers, M. F., Lokerson, J., Luchow, J., & Horvath, M. (1993). The devel-opment of a competency list for teachers of students with learning disabilities. Learning Disabilities Research & Practice, 8, 188–199.

Greenwood, J. (1984). My anxieties about math anxiety. Mathematics Teacher, 77, 662–663.

Grise, P. (1980). Florida's minimum competency testing for handicapped students. Exceptional Children, 47, 186–191.

Kamii, C., & Lewis, K. (1993). Primary arithmetic: Children inventing their own pro-cedures. Arithmetic Teacher, 41(4), 200–203.

Koscinski, S. T., & Hoy, C. (1993). Teaching multiplication facts to students with learn-ing disabilities: The promise of constant time delay procedures. Learning Disabilities: Research and Practice, 8(4), 260–263.

Lane, S. (1993). The conceptual framework for the development of a mathematics performance assessment. Educational Measurement: Issues and Practice, 12(2), 16–23.

Lepore, A. (1979). A comparison of computational errors between educable mentally retarded and learning disabled children. Focus on Learning Problems in Mathematics, 1(1), 12-33.

Lucas-Fusco, L. M. (1993). A content analysis of course syllabi in special education teacher preparation. Dissertation Abstracts International (UMI No. AAI9330093).

Mastropieri, M. A., Scruggs, T. E., & Shiah, S. (1991). Mathematics instruction for learning disabled students: A review of literature. Learning Disabilities Research and Practice, 6(2), 14–19.

Miller, S. P., & Mercer, C. D. (1993). Using data to learn about concrete–semiconcrete–abstract instruction for students with math disabilities. Learning Disabilities Research and Practice, 8(2), 89–96.

Montessori, M. (1964). The Montessori method. New York: Schocken Books.

National Assessment of Educational Progress. (1992). Can students do mathematical problem solving? Princeton, NJ: Educational Testing Service.

National Council of Teachers of Mathematics. (1989). Curriculum and evaluation stan-dards for school mathematics. Reston, VA: Author.

National Council of Teachers of Mathematics. (1991). Professional standards for teach-ing mathematics. Reston, VA: Author.

National Council of Teachers of Mathematics. (1993). Assessment standards for teaching mathematics. Reston, VA: Author.

O'Neill, J. (1988). How "special" should the special education curriculum be? Curriculum Update Monograph. Alexandria, VA: Association for Supervision and Curriculum Development.

Parmar, R. S., & Cawley, J. F. (1995). *Mathematics curricula frameworks: Goals for general and special education. Focus on Learning Problems in Mathematics, 17(5),* 50–66.

Parmar, R. S., Cawley, J. F., & Frazita, R. R. (1993). *Special project: Training program in mathematics. Unpublished manuscript, State University of New York at Buffalo.*

Parmar, R. S., Frazita, R. R., & Cawley, J. F. (1996). *Word problem–solving by students with mild disabilities and normally achieving students. Exceptional Children, 62, 415–430.*

Parmar, R. S., Klenk, L., & Cawley, J. F. (1996). *Research in academic skill areas for students with mild disabilities. Unpublished manuscript, State University of New York at Buffalo.*

Pelosi, P. (1977). *A report on computational variations among secondary school students. Unpublished manuscript, University of Connecticut.*

Piaget, J. (1965). *The child's conception of number. New York: W.W. Norton.*

Piccillo, B. (1994). *Science instruction for students with disabilities in included and non-included settings. Unpublished doctoral dissertation, State University of New York at Buffalo.*

Pieper, E., & Deshler, D. (1980). *A comparison of learning disabled adolescents with specific arithemtic and reading difficulties. (ERIC Document Reproduction Service No. ED 217 639)*

Resnick, L. B., & Ford, W. W. (1981). *The psychology of mathematics instruction. Hillsdale, NJ: Erlbaum.*

Riley, M. S., Greeno, J. G., & Heller, J. I. (1983). *Development of children's problem-solving ability in arithmetic. In H. P. Ginsburg (Ed.), The development of mathematical thinking (pp. 153–196). New York: Academic Press.*

Rourke, B., & Strang, J. D. (1983). *Subtypes of reading and arithmetic disabilities: A neuropsychological analysis. In. M. Rutter (Ed.), Developmental neuropsychiatry (pp. 477–488). New York: Guilford.*

Senio, P., Urban, A., Bukowski, J., Achenbach, K., Buchnowski, M., Smeal, T., Argenio, S., & DeLuca, C. (1995, November). *Using alternative mathematics assessments in general and specialeducation classrooms. Paper presented at the annual meeting of the New York State Council for Exceptional Children, Niagara Falls, NY.*

Skinner, B. F. (1968). *The technology of teaching. New York: Appleton-Century-Crofts.*

Thorndike, E. L. (1906). *The principles of teaching based on psychology. New York: Seiler.*

Unglab, K. W. (1995). *Mathematics anxiety in preservice elementary school teachers. Unpublished doctoral dissertation, State University of New York at Buffalo.*

Warner, M., Alley, G., Schumaker, J., Deshler, D., & Clark, F. (1980). *An epidemiological study of learning disabled adolescents in secondary schools: Achievement and ability, socioeconomic status and school experiences (Research Report No. 13). Lawrence: Institute of Research in Learning Disabilities, University of Kansas.*

Webster, R. E. (1980). *Short-term memory in mathematics deficient students as a function of input-output modalities. The Journal of Special Education, 14, 67–78.*

CHAPTER 12

Ackerman, P. T., Anhalt, J. M., & Dykman, R. A. (1986). Arithmetic automatization failure in children with attention and reading disorders: Associations and sequelae. Journal of Learning Disabilities, 19, 222–232.

Anderson, J. R. (1982). Acquisition of cognitive skill. Psychological Review, 89, 369–406.

Anderson, J. R. (1987). Skill acquisition: Compilation of weak-method problem solutions. Psychological Review, 94, 192–210.

Babbitt, B. C., & Miller, S. P. (1996). Using hypermedia to improve the mathematics problem-solving skills of students with learning disabilities. Journal of Learning Disabilities, 29, 391–401, 412.

Barron, B., Vye, N. J., Zech, L., Schwartz, D., Bransford, J. D., Goldman, S. R., Pellegrino, J., Morris, J., Garrison, S., & Kantor, R. (1995). Creating contexts for community-based problem solving: The Jasper Challenge Series. In C. N. Hedley, P. Antonacci, & M. Rabinowitz (Eds.), Thinking and literacy: The mind at work (pp. 47–71). Hillsdale, NJ: Erlbaum.

Blumenfeld, P. C., Soloway, E., Marx, R. W., Krajcik, J. S., Guzdial, M., & Palincsar, A. (1991). Motivating project-based learning: Sustaining the doing, supporting the learning. Educational Psychologist, 26, 369–398.

Bottge, B. A., & Hasselbring, T. S. (1993). A comparison of two approaches for teaching complex, authentic mathematics problems to adolescents with learning difficulties. Exceptional Children, 59, 556–566.

Bransford, J. D., Franks, J. J., Morris, C. D., & Stein, B. S. (1979). Some general constraints on learning and research. In L. S. Cermak & F. I. M. Craik (Eds.), Levels of processing and human memory (pp. 331–354). Hillsdale, NJ: Erlbaum.

Bransford, J. D., Goldman, S. R., & Hasselbring, T. S. (1995, April). Marrying constructivist and skills-based approaches: Could we, should we, and can technology help? Symposium presented at the annual meeting of the American Educational Research Association, San Francisco.

Bransford, J. D., Goldman, S. R., & Vye, N. J. (1991). Making a difference in people's abilities to think: Reflections on a decade of work and some hopes for the future. In L. Okagaki & R. J. Sternberg (Eds.), Directors of development: Influences on children (pp. 147–180). Hillsdale, NJ: Erlbaum.

Bransford, J. D., Sherwood, R. S., Vye, N. J., & Rieser, J. (1986). Teaching thinking and problem solving: Research foundations. American Psychologist, 41, 1078–1089.

Bransford, J. D., Sharp, D. M., Vye, N. J., Goldman, S. R., Hasselbring, T. S., Goin, L., O'Banion, K., Livernois, J., Saul, E., & the Cognition and Technology Group at Vanderbilt. (1996). MOST environments for accelerating literacy development. In S. Vosniadou, E. De Corte, R. Glaser, & H. Mandl (Eds.), International perspectives on the psychological foundations of technology-based learning environments (pp. 223–256). Hillsdale, NJ: Erlbaum.

Brown, A. L., & Campione, J. C. (1990). Communities of learning and thinking or a context by any other name. Human Development, 21, 108–125.

Brown, A. L., & Campione, J. C. (1994). *Guided discovery in a community of learners.* In K. McGilly (Ed.), *Classroom lessons: Integrating cognitive theory and classroom practice* (pp. 229–272). Cambridge, MA: MIT Press.

Brown, C., Carpenter, T., Kouba, V., Lindquist, M., Silver, E., & Swafford, J. (1988a, April). *Secondary school results for the fourth NAEP assessment: Discrete mathematics, data organization and interpretation, measurement, number and operations. Mathematics Teachers, 81,* 241–248.

Brown, C., Carpenter, T., Kouba, V., Lindquist, M., Silver, E., & Swafford, J. (1988b). *Secondary school results for the fourth NAEP assessment: Algebra, geometry, mathematical methods and attitudes. Mathematics Teachers, 81,* 337–347, 397.

Bruner, J. (1960). *The process of education: A landmark in educational theory.* Cambridge, MA: Harvard University.

Carnine, D., Jones, E., & Dixon, R. (1994). *Mathematics: Educational tools for diverse learners. School Psychology Review, 23,* 406–427.

Cawley, J. F., & Miller, J. H. (1989). *Cross-sectional comparisons of the mathematical performance of children with learning disabilities: Are we on the right track toward comprehensive programming? Journal of Learning Disabilities, 23,* 250–254, 259.

Charles R., & Silver, E. A. (Eds.). (1988). *The teaching and assessing of mathematical problem solving.* Hillsdale, NJ: Erlbaum.

Cognition and Technology Group at Vanderbilt. (1990). *Anchored instruction and its relationship to situated cognition. Educational Researcher, 19(6),* 2–10.

Cognition and Technology Group at Vanderbilt. (1992a). *An anchored instruction approach to cognitive skills acquisition and intelligent tutoring.* In J. W. Regian & V. J. Shute (Eds.), *Cognitive approaches to automated instruction* (pp. 135–170). Hillsdale, NJ: Erlbaum.

Cognition and Technology Group at Vanderbilt. (1992b). *The Jasper experiment: An exploration of issues in learning and instructional design. Educational Technology Research and Development, 40,* 65–80.

Cognition and Technology Group at Vanderbilt. (1992c). *The Jasper series as an example of anchored instruction: Theory, program description, and assessment data. Educational Psychologist, 27,* 291–315.

Cognition and Technology Group at Vanderbilt. (1993a). *Anchored instruction and situated cognition revisited. Educational Technology, 33,* 52–70.

Cognition and Technology Group at Vanderbilt. (1993b). *The Jasper series: Theoretical foundations and data on problem solving and transfer.* In L. A. Penner, G. M. Batsche, H. M. Knoff, & D. L. Nelson (Eds.), *The challenges in mathematics and science education: Psychology's response* (pp. 113–152). Washington, DC: American Psychological Association.

Cognition and Technology Group at Vanderbilt. (1994a). *From visual word problems to learning communities: Changing conceptions of cognitive research.* In K. McGilly (Ed.), *Classroom lessons: Integrating cognitive theory and classroom practice* (pp. 157–200). Cambridge, MA: MIT Press.

Cognition and Technology Group at Vanderbilt. (1994b). *Multimedia environments for developing literacy in at-risk students.* In B. Means (Ed.), *Technology and educational reform: The reality behind the promise* (pp. 23–56). San Francisco: Jossey-Bass.

Cognition and Technology Group at Vanderbilt. (1996a). Looking at technology in context: A framework for understanding technology and education. In D. C. Berliner & R. C. Calfee (Eds.), The handbook of educational psychology (pp. 807–840). New York: Macmillan.

Cognition and Technology Group at Vanderbilt. (1996b). Multimedia environments for enhancing learning in mathematics. In S. Vosniadou, E. de Corte, R. Glaser, & H. Mandl (Eds.), International perspectives on the psychological foundations of technology-based learning environments (pp. 285–306). Hillsdale, NJ: Erlbaum.

Cognition and Technology Group at Vanderbilt. (in press-a). The Jasper project: Lessons in curriculum, instruction, assessment, and professional development. Mahwah, NJ: Erlbaum.

Cognition and Technology Group at Vanderbilt. (in press-b). The Jasper series: A design experiment in complex, mathematical problem-solving. In J. Hawkins & A. Collins (Eds.), Design experiments: Integrating technologies into schools. New York: Cambridge University Press.

Collins, A., Hawkins, J., & Carver, S. M. (1991). A cognitive apprenticeship for disadvantaged students. In B. Means, C. Chelemer, & M. S. Knapp (Eds.), Teaching advanced skills to at-risk students (pp. 216–243). San Francisco: Jossey-Bass.

Fleischner, J. E., Garnett, K., & Shepherd, M. (1982). Proficiency in arithmetic basic fact computation by learning disabled and nondisabled children. Focus on Learning Problems in Mathematics, 4, 47–55.

Fuchs, L. S., Fuchs, D., & Hamlett, C. L. (1992). Computer applications to facilitate curriculum-based measurement. Teaching Exceptional Children, 24(4), 58–60.

Fuchs, L. S., Fuchs, D., Hamlett, C. L., & Stecker, P. M. (1991). Effects of curriculum-based measurement on teacher planning and student achievement in mathematics operations. American Educational Research Journal, 28, 617–641.

Gagné, R. M. (1968). Contributions of learning to human development. Psychological Review, 75, 177–191.

Ginsburg, H. (1977). Children's arithmetic: The learning process. New York: D. Van Nostrand.

Goldman, S. R., Mertz, D. L., & Pellegrino, J. W. (1989). Individual differences in extended practice functions and solution strategies for basic addition facts. Journal of Educational Psychology, 81, 481–496.

Goldman, S. R., & Pellegrino, J. W. (1986). Effective drill and practice on the microcomputer. Academic Therapy, 22, 133–140.

Goldman, S. R., & Pellegrino, J. W. (1987). Information processing and educational microcomputer technology: Where do we go from here? Journal of Learning Disabilities, 20, 144–154.

Goldman, S. R. , Pellegrino, J. W., & Mertz, D. L. (1988). Extended practice of basic addition facts: Strategy changes in learning disabled students. Cognition & Instruction, 5, 223–265.

Graham, S., & Harris, K. (1989). Improving learning disabled students' skills at composing essays: Self-instructional strategy training. Exceptional Children, 56, 201–214.

Graham, S., & Harris, K. (1994). Implications of constructivism for teaching writing to students with special needs. The Journal of Special Education, 28, 275–289.

Hasselbring, T. S., Goin, L. I., Alcantara, P., & Bransford, J. D. (1991, April). Developing mathematical fluency in children with learning handicaps. Paper presented at the annual meeting of the American Educational Research Association, Boston.

Hasselbring, T. S., Goin, L., & Bransford, J. D. (1988). Developing math automaticity in learning handicapped children: The role of computerized drill and practice. Focus on Exceptional Children, 20(6), 1–7.

Hasselbring, T. S., Sherwood, R. D., Bransford, J. D., Mertz, J., Estes, B., Marsh, J., & Van Haneghan, J. (1991). An evaluation of specific videodisc courseware on student learning in a rural school environment (Technical Report). Nashville, TN: Vanderbilt University, Learning Technology Center.

Hiebert, J., & Lefevre, P. (1986). Conceptual and procedural knowledge in mathematics: An introductory analysis. In J. Hiebert (Ed.), Conceptual and procedural knowledge: The case of mathematics (pp. 1–27). Hillsdale, NJ: Erlbaum.

Hofmeister, A. M. (1993). Elitism and reform in school mathematics. Remedial and Special Education, 14(6), 8–13.

Kouba, V. L., Brown, C. A., Carpenter, T. P., Lindquist, M. M., Silver, E. A., & Swafford, J. O. (1988). Results of the fourth NAEP assessment of mathematics: Number, operations, and word problems. Arithmetic Teacher, 35(8), 14–19.

Lesh, R. (1981). Applied mathematical problem solving. Educational Studies in Mathematics, 12, 235–264.

Mercer, C. D., Harris, C. A., & Miller, S. P. (1993). Reforming reforms in mathematics. Remedial and Special Education, 14(6), 14–19.

Morris, C. D., Bransford, J. D., & Franks, J. J. (1979). Levels of processing versus transfer appropriate processing. Journal of Verbal Learning and Verbal Behavior, 16, 519–533.

National Council of Teachers of Mathematics. (1989). Curriculum and evaluation standards for school mathematics. Reston, VA: Author.

New Standards. (1995). New Standards portfolio field trial: 1995–96 workshop edition, high school. Rochester, NY: Author.

Nickerson, R. S. (1988). On improving thinking through instruction. Review of Research in Education, 15, 3–57.

Pellegrino, J. W., & Goldman, S. R. (1987). Information processing and elementary mathematics. Journal of Learning Disabilities, 20, 23–32, 57.

Pellegrino, J. W., Hickey, D., Heath, A., Rewey, K., Vye, N. J., & Cognition and Technology Group at Vanderbilt. (1991). Assessing the outcomes of an innovative instructional program: The 1990–1991 implementation of the "Adventures of Jasper Woodbury" (Tech. Rep. No. 91-1). Nashville, TN: Vanderbilt University, Learning Technology Center.

Peterson, P. L., Fennema, E., & Carpenter, T. (1991). Using children's mathematical knowledge. In B. Means, C. Chelemer, & M. S. Knapp (Eds.), Teaching advanced skills to at-risk students (pp. 68–111). San Francisco: Jossey-Bass.

Porter, A. (1989). A curriculum out of balance: The case of elementary school mathematics. Educational Researcher, 18, 9–15.

Resnick, L. (1987). Education and learning to think. Washington, DC: National Academy Press.

Resnick, L. B., Bill, V. L., Lesgold, S. B., & Leer, M. N. (1991). Thinking in arithmetic class. In B. Means, C. Chelemer, & M. S. Knapp (Eds.), Teaching advanced skills to at-risk students (pp. 27–53). San Francisco: Jossey-Bass.

Resnick, L. B., & Klopfer, L. E. (Eds.). (1989). Toward the thinking curriculum: Current cognitive research. Alexandria, VA: Association for Supervision and Curriculum Development.

Rivera, D. (1993). Examining mathematics reform and the implications for students with mathematical disabilities. Remedial and Special Education, 14(6), 24–27.

Russell, R. L., & Ginsburg, H. P. (1984). Cognitive analysis of children's mathematics difficulties. Cognition & Instruction, 1, 217–244.

Scardamalia, M., & Bereiter, C. (1991). Higher levels of agency for children in knowledge building: A challenge for the design of new knowledge media. Journal of the Learning Sciences, 1, 37–68.

Schoenfeld, A. H. (1989). Teaching mathematical thinking and problem solving. In L. B. Resnick & L. E. Klopfer (Eds.), Toward the thinking curriculum: Current cognitive research (pp. 83–103). Alexandria, VA: ASCD.

Silver, E. A. (1986). Using conceptual and procedural knowledge: A focus on relationships. In J. Hiebert (Ed.), Conceptual and procedural knowledge: The case of mathematics (pp. 181–189). Hillsdale, NJ: Erlbaum.

U.S. Department of Labor. (1992). Secretary's Commission on Achieving Necessary Skills report for America 2000. Washington, DC: Author.

Vye, N. J., Sharp, D. M., McCabe, K., & Bransford, J. D. (1991). Commentary: Thinking in arithmetic class. In B. Means, C. Chelemer, & M. S. Knapp (Eds.), Teaching advanced skills to at-risk students (pp. 54–67). San Francisco: Jossey-Bass.

Whitehead, A. N. (1929). The aims of education. New York: Macmillan.

Contributors

Diane S. Bassett, faculty member, University of Northern Colorado

Shalini Bhojwani, master's degree student, special education, Bowling Green State University

Brian R. Bryant, private consultant, Austin, Texas

Douglas Carnine, professor, education, University of Oregon, and director, National Center to Improve the Tools of Educators

John F. Cawley, professor, special education, State University of New York at Buffalo

Cognition and Technology Group at Vanderbilt, members: Brigid Barron, John Bransford, Trevor Davies, Laura Goin, Allison Moore, Priscilla Moore, Tom Noser, James Pellegrino, Daniel Schwartz, Nancy Vye, and Linda Zech

James A. Conway, doctoral candidate, clinical neuropsychology, University of Windsor

Mary E. Cronin, professor, special education and rehabilitative services, University of New Orleans

Herbert P. Ginsburg, professor, psychology and mathematics education, Columbia University

Susan R. Goldman, professor, psychology, and co-director, Learning Technology Center, Vanderbilt University

Ted S. Hasselbring, professor, special education, and co-director, Learning Technology Center, Vanderbilt University

Eric D. Jones, professor, special education, Bowling Green State University

Graham A. Jones, visiting professor, mathematics, Illinois State University

Annie E. Koppel, marketing assistant, Western Psychological Corporation

Cynthia W. Langrall, assistant professor, mathematics, Illinois State University

Cecil D. Mercer, professor, special education, University of Florida

Susan Peterson Miller, associate professor, special education, University of Nevada at Las Vegas

Marjorie Montague, professor, special education, University of Miami

Rene S. Parmar, associate professor, measurement and evaluation, St. John's University

James R. Patton, executive editor, PRO-ED

Diane Pedrotty Rivera, assistant professor, special education, University of Texas at Austin

Byron P. Rourke, professor, psychology, University of Windsor, and faculty member, Yale University

Carol A. Thornton, professor, mathematics, Illinois State University

Rich Wilson, professor, special education, Bowling Green State University

Author Index

Subject Index